In Search of the Green-Eyed Yellow Idol

About the authors

Siân and Bob were born in the years just after the war. Anyone of the same vintage will know which war we are talking about; there were not so many in those days. At the time rationing was still in place, and we recently discovered baby Bob's ration book, in which his weekly allowance of Ovaltine was recorded.

They both have a Senior Railcard, but neither of them is quite old enough – yet – to be eligible to draw a state pension.

They now spend most of their time between Chichester, Chamonix and Kathmandu.

Contact
sianpj@hotmail.com

Other books by the authors

Bradt
Africa Overland --- 2005, 2009, 2014

Cicerone
The Mount Kailash Trek --- 2007
Annapurna: A Trekker's Guide --- 2013

CreateSpace
All HMH titles below
Earthquake Diaries: Nepal 2015 --- 2015

Himalayan Map House (HMH)
Trekking around Manaslu and the Tsum Valley --- 2013
Dolpo --- 2014
Ganesh Himal --- 2014
Langtang --- 2014
Everest --- 2014
Rolwaling --- 2015
Nepal Himalaya --- 2015

Pilgrims
Kathmandu: Valley of the Green-Eyed Yellow Idol --- 2005
Ladakh: Land of Magical Monasteries --- 2006
Kailash & Guge: Land of the Tantric Mountain --- 2006

In Search of the Green-Eyed Yellow Idol

Around the World in Forty Years
Overland, Adventure & Trekking

Siân Pritchard-Jones
Bob Gibbons

© Siân Pritchard-Jones and Bob Gibbons

All rights reserved. No part of this publication may be reproduced or transmitted in any form or by any means without prior written permission of the copyright holders.

ISBN-13: 978-1519245380
ISBN-10: 1519245386

First edition: 18 September 2015

Published by Expedition World
www.expeditionworld.com
email: sianpj@hotmail.com

Photographs

Above: Typical idols in Kathmandu
Cover: Our Land Rover in Lhasa, Tibet
Back Cover: Authors at Dallol, Danakil, Ethiopia
Title page: Pushing on to the Depression du Mourdi, Chad

Contents

About the authors..2
Contact..2
Other books by the authors...2
Acknowledgements...7
Preface..8
Introduction 10
PART ONE: A PIG FARMER'S SON – BOB 12
Oh Calamity – Overland 1974...13
Back on the road: January 1975..33
Trans-African Ordeals: 1975–76..42
Trans-Africa again: 1976...52
The Exodus Years 54
South America: 1977 57
Back to the Exodus fold: 1978 61
Nepal-bound 69
The Iranian Revolution: 1979..70
A proper job! 72
Sudan or bust..75
South America Overland 77
Another winter in Nepal...82
PART TWO: A COUPLE FOR THE ROAD..85
Siân... from Moel Siabod through Oxford to Nepal...................................85
Addicted to India: 1984 90
Delhi to Kathmandu 95
Interlude in Langtang: November 1984...104
Lhasa to Kashgar: February 1985 106
Trans-Siberian Railway: March 1985...122
Put on your trekking boots 125
The Alps: Summer 1985 125
Back to Kathmandu when all else fails: Autumn 1985 125
On the roof of the world at last: November 1985 125
A couple for the road...129
District Commissioner's Office, Kathmandu: 16 January 1986...............129
A window on Yemen: March 1986 130
Chamonix: a pied-a-terre in 1986 134
Unemployed and re-employed again 135
Facets of West Africa: 1990 135
Kathmandu Under Curfew: April 1990 138
Rheumatoid Arthritis onset: July 1991 144
Mayan Odyssey: March 1992..144
Mustang: September 1992..147
Mind Your Step In Cambodia: November 1993 148
Vietnam & Laos...154
The Pilgrims Way..158
Those overland buses again...160
Middle East: 1994 160
Asia Overland trials: 1995–96...162
Cuba, Dominican Republic and Haiti: 1997–8 168
PART THREE: THE NEW MILLENNIUM 170
Overland for Oldies, No Problem: October 1999 170
Interlude in Bhutan: January 2000 177
The Dark Age of Kali: 2000–1 180

Militant tendencies: 2001–2 182
The Caucasus unlocked: 2001 185
Nomads of the Sahara 194
Mauritania: 1993 194
The Ténéré of Niger: October 2000 197
Algeria, the Tassili Plateau at last: April 2001.....................................204
Chad: 2002 213
Libyan Journey: 2002..224
Libreville or bust: 2003..229
Central Asia: 2003 239
London to Cape Town: 2004..250
Travellers' Tales..286
Island Paradises: 2004–5 286
Maldives: December 2004 286
South America: February–April 2005..287
Cape Verde: May 2005...291
Mount Kailash and Tsaparang: October 2005 292
More guidebooks 294
Ladakh and Spiti: February 2006..294
Canyons and Deserts: April–May 2006 296
Korean surprises: December 2006 296
Karakoram Highway: October 2007..299
The Overland Bug Bites Again..303
West Africa Overland: January–April 2008 303
Caribbean cruising: December 2008 312
Eastern Europe Overland: January–March 2009................................313
Indian Ocean island delights: May 2009...314
Northern South America backwaters: January 2010 317
Africa Overland: 2010 319
Siân's sad days: May 2010 327
A boiling Indian Odyssey: May 2010...327
Bandits, paradise and a canal: January 2011......................................328
Asia Overland through Tibet: 2011 330
Another dream, another adventure 343
Off the map in Somaliland: January 2014..343
Into the Cauldron of Fire, Erta Ale: January 2014...............................345
Down Under briefly: Spring 2015...348
Earthquakes in Nepal: April – May 2015...351
In Search of the Green-Eyed Yellow Idol...353

Acknowledgements

Special thanks to our parents, Beryl & Tony Gibbons and Marianne & John Pritchard-Jones, for their tacit support throughout all those years of avoiding a proper job. And to Bob's uncles John and Mick Gibbons, for their help preparing the first Land Rover.

Our oldies' bus in Gorakhpur, India 2000

Thanks to Sherpa Expeditions for employing us every summer for over 30 years and making all this possible. Also to Trans-African Expeditions, Exodus Travels, Hann Overland, Australian Himalayan Expeditions, Ama Dablam Trekking, World Expeditions, Bradt Travel Guides, Cicerone Press, Pilgrims Book House and Himalayan Map House.

Thanks to David Durkan in Kathmandu, and Kev Reynolds in the Kentish Alps, for their helpful comments on the first draft.

And to all our friends and fellow travellers we have met on our journeys...

Preface

It's a dangerous business, Frodo, going out your door. You step on to the road, and if you don't keep your feet, there's no knowing where you might be swept off to.
The Lord of the Rings, J.R.R. Tolkien

Boudhanath Stupa, Kathmandu

There's a green-eyed yellow idol to the north of Kathmandu…

This is the first line of a poem by J. Milton Hayes that inspired this narrative.

As the veil of early morning mist lifts from the Kathmandu Valley, the all-seeing eyes of the great stupa gaze out across a crisp, mellow sky. The eyes stare with benign passion, never blinking. On the northern horizon the glittering peaks of Ganesh Himal and Gauri Shankar glow red as another day dawns in the Himalaya. We are in a rooftop café at Boudhanath Stupa outside Kathmandu. The powerful attraction of the mystical eyes and the fabled idols are a captivating sight, too mysterious and abstract to comprehend.

In this life everyone is on a journey, even though some never stray far from home. Ours is a journey that has sometimes been more enchanting in the planning than in the execution, its direction apparently chosen by some unknown compulsion. This unknown force still drives us constantly 'around the next bend', time after time.

This narrative is drawn from a collection of letters, memoirs and diaries since 1974. It is by any definition a collection of self-centred stories, for how else can we portray the life we have lived, a life so often detached from the 'normal' everyday path? We have been fortunate to be able to follow our dreams with such insatiable passion.

During moments of quiet reflection, we may all search for a reason for what we do in life. As we drift or are driven through life, do we follow any innate sense of purpose?

To travel is to uncover more questions than answers. Do we understand more about the ways of the world through our travels, or do we clog up our minds with sights that underscore the fog or 'futility' of humanity?

Will we be searching forever for the elusive green-eyed yellow idol, a symbol of the mysteries of life itself?

Siân and Bob
Kathmandu 2015

A mandala of coloured powder in Bungamati, Kathmandu

The impermanence of everything
Bungamati is home to the Red Machhendranath, the rain god of Kathmandu. The idol is carried between Patan and Bungamati during a spring festival on a tall, rickety wooden chariot. After the celebrations the mandala is left to blow away in the wind when nature decrees, hinting at the temporary nature of everything, whether man or mountain, in this universe.

Introduction

Not all those who wander are lost.
Lord of the Rings, J.R.R. Tolkien

From 1974 Bob drove overland trucks across Asia, Africa and South America. Siân had a more conventional start, with a career in computing but a passion for travel. In 1983 we met in the Himalaya (of course!) and married in Kathmandu in 1986. Since then we have worked as trekking guides in the Alps and the Himalaya, also driving tours and overland trips.

This book is a collection of stories, all based on true events that we have experienced, both good and bad. Along the way we attempt to provide some insights into the places we have visited. Inevitably there are questions about the destinations, the cultures and the experiences on our journeys. We are lucky to have been born at a time when we could travel through these countries, so many of which are now too dangerous to contemplate. Certainly it is not as easy as it may seem at first sight to reject the safer, probably more comfortable life. We have often survived on bread and bananas, sleeping on the ground or the roof of the Land Rover, or in a dodgy hotel with a cupboard jammed against the door.

In many ways this narrative is a window on the last forty years. We cannot but help consider some of the historical and political background of the places visited and question the impact of such events on the people around us. Looking around the modern world, people seem to be increasingly in a frenetic hurry, focussed in some direction or on some ambition. For many other people, simply surviving is a struggle.

We have witnessed the phenomenal growth in the size of most cities across the world: Istanbul, Delhi, Kathmandu, Lhasa, Beijing, Cairo, Damascus, Nairobi, Sao Paulo and La Paz for starters. As the world's population grows, some may say catastrophically, everyone clamours for his or her own space and necessary resources. Places we remember fondly have become engulfed by wars, rebellions and natural disasters. Often there is no apparent reason other than human stubbornness why a quarrel develops into a war, with neither side able to compromise before taking to arms.

Sitting for hours on a hot stuffy train, or crammed into seats designed for midgets in a dangerously overloaded bus, travellers have to put their faith in destiny. With hours to while away, it's easy to ponder on life's unfair challenges and its mismatch of random gifts. In observing the passing life of the road, we are bound to question the reasons for so much suffering, intolerance and aggression. We sometimes despair about the depths to which humanity is sinking and cannot understand why it happens. Yet there is inspiration everywhere, especially in the qualities of the human spirit, the overwhelming vibrancy of people. As the bus bumps along for an eternity, we ask no more than to count our blessings.

Any narrative comes from a personal point of view; we never really know if we are thinking in the same way as our neighbours, whoever they may be. Does anyone ever really understand other people, especially those from different cultures? How much do people make their own destiny, wherever they may be? Do the people we meet along the way bring us inspiration or depression? Even a half-hearted attempt to understand how other cultures function, what the people aspire to, how they view us and interact with us, might lead to a more harmonious world.

The problems of the modern world are mammoth, driven by ego, power, greed, manipulation, control of others and the age-old conundrum of the question what is life itself. Despite so much innovative communications technology, we still seem no closer to understanding each other. After more than forty years of travelling, living and working with foreign cultures, there seem to be more angles, more questions and less certainty about anything.

Several years ago we started working occasionally for Pilgrims Book House in Thamel, Kathmandu, editing and writing back covers for books that were out of print in the west. We then moved on to a series of trekking guidebooks for Himalayan Map House, as well as Cicerone and Bradt in the UK. Writing back covers and prefaces, then practical guidebooks, seemed easy compared to embarking on an autobiographical collection.

For years now it's been said to us, 'Why don't you write a book about your experiences?' But writing a book about travelling is one thing; to make it relevant and interesting to others is quite another matter. Most of the text relates to the situation at the time of the experience, whether written with the naivety of youth or the cynicism of imminent old age. It is not all true today in 2015.

This book started literally as an attempt to put pen to paper almost before computers were invented! It's a travelogue, laced with historical comment and the experiences of our time on the road that may offer some ideas to future travellers. It's the sort of book that can't be read in full flow without some recuperative lulls (of days or weeks!). Delve into it when your mood invites! With the way the world is 'progressing', it may appeal more to armchair travellers, that is if their armchair upholstery satisfies current fire regulations.

We have attempted to bring the stories to life in the same way that the experiences were lived. It may fail badly, like any venture into the unknown. It is sometimes critical of the establishment, who seem to be forever trying to control and limit what we can do, even if they have the best of intentions. The regimented, computer-controlled modern world is a hard place for modern (and traditional) nomads to co-exist. This narrative certainly has a pinch of philosophy induced by those wild places where the silence of the night is deafening under a starry sky. In truth it is inspired by all the wonders of the world, whether natural or man-made, and by many of the people we have met along the way.

Sometimes you find yourself in the middle of nowhere, and sometimes, in the middle of nowhere, you find yourself
Unknown

Salt lakes and flamingos in southern Bolivia

PART ONE: A PIG FARMER'S SON – BOB

As a child from a farming family, I lived the first five years of life in relative isolation from other children. We were only a few miles from the town, but we lived in our own world of dreams, my sister and I. We invented imaginary lands of magic, a dreamland where we empowered our simple toys with unimaginable powers of wizardry and daring. We had so few toys we were forever looking for new incarnations of the same toys. Emerging from this isolated background seems to have given me a desire to seek out new horizons. Just as a mountaineer is drawn to conquer preposterous new peaks, so I have felt compelled to explore the world.

For as long as I can remember, I spent hours and hours after school just staring at maps and atlases. I still do. I can rattle off endless place names and always there is some new horizon to plan for. This desire for exploration is of course not in anyway unique. It can certainly land you in hot water; things move on at home in your absence, picking up temporary or seasonal work became harder after 1980. You come back with new ideas, often restless, usually broke and perhaps with a feeling of alienation from normal daily life. It's easy to feel different and apart. It is not a conscious thing, but it creeps up on you like a demon that suddenly looms out of the shadows.

In the great adventure of youth, I, like others, was so naïve. The following excerpts from letters sent back on that first journey will testify to a complete lack of understanding and innocence, but so much ignorance as well. I couldn't wait to leave college. I dutifully went through the motions of study and attempted to immerse myself in the subjects. In reality it was a charade. I studied Engineering Geology, thinking that this was a passport for a job travelling the world at someone else's expense. It was nothing of the sort. My fellow students invariably ended up in far-flung corners of the world, but they also ended up stuck in some desert compound or drilling holes in murky backwaters beside a motorway as the rain poured down. Of course some enjoyed all that and some made lots of money in Arabia. But this was not for me. I just had to go overland to Kathmandu and see the mysterious eyes of the golden stupa. Those same eyes that had stared out at me from the pages of a book called 'A Winter in Nepal' (by John Morris), which I had found in the local library. Those same eyes that had made me dream about the land of the Himalaya, Nepal. Never mind the Geology.

An outsider may well think I was some crazed hippy frequently 'off my trolley', and that indeed may be so for other reasons. At college I was quiet and reserved I could never have been a real hippy. I could never have been even 'a plastic hippy' dabbling in mild drugs, curing the world of all ills but not going totally off the rails. The plastic hippies were often those who partied in Goa over Christmas and then became lawyers and barristers back home. What happened to those characters?

The Nepalese men contemplated the ills of the world under a fragile, overhanging temple or played music after sunset. The women were goddesses in our lusting youthful eyes, but their dignity overpowered us amid these scenes of western decadence. We might have felt superior, sanctimonious and self-righteous in those freaking streets, but secretly we also wondered if we might be missing out on some alluring or erotic experience of the mind by rejecting the offers of hash and heavier hallucinating substances. We should not be so critical now, forty years later. Life is complex; nothing is ever as it seems, contradictions and morality confront us daily. It's hard to know what is truth; there is no true black or white, only endless shades of grey.

The hippies of Kathmandu
Back in 1974, though, we met the real hippies, the lost children who found ganga and 'paradise man'… who found a spiritual awakening in the dung-infested streets behind Hanuman Dhoka palace in old Kathmandu. But they couldn't work out how to get an exit visa from the intoxicating lifestyle. I can't imagine what our parents thought would happen to us, or if we would ever come back sane. We were three naïve lads looking for adventure. Our preconceived ideas about the enchanting valley of Kathmandu took a bit of a battering as we witnessed the absurdity of western dropouts completely obliterating their last vestiges of intelligence in the murky corners of Freak Street. One tragic Italian hippy died in Freak Street, a helpless shell of humanity; a truly pathetic sight amid the genuine poverty of the Nepalese.

Maybe we were just a bit envious of the hippies, who had no voices in their minds telling them to keep a clear head. Yet for all that decadence, I feel a sad nostalgia when I walk down Freak Street today, with its plush new shops. The Monumental Guesthouse is still there, as is the old low-roofed restaurant building that was once called Eat at Joe's, whose food constantly made us feel ill. But the old atmosphere is lost forever.

Of course the hippies were in a strange way the gift that brought the bounty to Nepal in subsequent years, so we can only thank them.

Freak Street, Kathmandu 1974

After the hippies came Thamel, the new trendier area of Kathmandu for travellers, a new Mecca for trekkers. The early years of Thamel were tranquil; the old wooden and brick houses slowly fell down and were replaced by less attractive but functional cafés, restaurants and hotels. The Kathmandu Guest House became the place to be. The precarious temple just down from KC's restaurant and Bambooze Bar finally collapsed, nudged on its way by an Encounter Overland truck squeezing into the confines of the 'Kathouse' – the Kathmandu Guest House. But Thamel was just an overgrown collection of walled lanes and traditional houses, including a couple of larger houses with gardens in 1974.

We were ready to embark on our big journey in May 1974, a journey born from a surfeit of imagination that would change us forever.

Oh Calamity – Overland 1974

1971
My sister tossed the magazine across the room and I spotted the relevant advert: 'Indigo Overland, Morocco 3 weeks, Overland to Kathmandu – 10 weeks £99.'

Later in the year I did the trip to Morocco. It was fascinating, especially the souks and markets of old Fez. I didn't like the group thing so much – lots of boozing and no flexibility. Would it be Kathmandu next?

Somehow it all came together in 1974 after college. Hugh, Pete and I bought an old Land Rover and fixed it up for the roads of Asia. After working to get the funds, we eventually set off in May, some 11 months after leaving Portsmouth Polytechnic. Not the best time of year.

Departure day 1974: Tony (Bob's father), Pete, Beryl (Bob's mother), sister Christine and Grandma Lillian

We wound our way around Europe, with brief visits to Hungary, Romania, Yugoslavia and Bulgaria. Eastern Europe was extremely backward, with terrible roads, poor housing and dodgy-looking characters hanging about on street corners. Greece was baking under the early summer sun. We camped wild among the sweet smells of thyme, sage and olives beneath rugged mountains and close to idyllic beaches. We trekked up Mount Olympus and got lost coming down. The package holiday crowd had hardly been invented at this time, so Greece was a revelation of tranquillity.

Camping wild in Greece 1974

Istanbul and the gates of Asia – what an exotic mix it was back then: the crazy water sellers, the spice bazaar, the colourful covered souks, the slender minarets and the Pudding Shop. This was the first mecca on the hippy trail, a haven of dense, acrid smoke cluttered with humanity saving the planet and immersed in self-deprecating indulgence. At that time the ancient walls virtually marked the edge of the city, unlike today where miles of suburbs stretch westwards almost to Greece.

We toured around the Turkish coast, soaking up the warm waters of the summer Mediterranean. The antiquities of Ephesus, Side and Aspendos led on to the crusader castle at Anamur. History unfolded daily, like some exotic story from the pages of an ancient book. Heading across the Tarsus Mountains we climbed up to Konya for the Mevlana whirling dervishes and on to fabled Cappadocia. The amazing volcanic structures of the region, the towering hollowed-out pumice towers, cave churches and deep underground troglodyte cities astonished us in equal measure. If this was life at full speed, it was more intoxicating than any drug.

Crossing into Syria, we were invited for dinner in Aleppo after stopping to ask the way. We shared a fabulous local dinner, sitting cross-legged on the floor around a series of magnificent silver family-sized platters. No one could converse, but the camaraderie of sharing was startling for us, coming from such a conservative society. The people across Syria were truly hospitable, as was the traditional welcome of Islam. Lebanon had not yet been riven by its long civil war, so we stopped in cosmopolitan Beirut and met an Indigo Overland group trying to get visas for Iraq. We all failed with Iraq, but somehow got Saudi transit visas.

Baalbek remains, Lebanon 1974

After crossing the mountains of Lebanon, we arrived in Damascus, stepping back into biblical times in the grand bazaar. In Jordan the ruins of Jerash enthused us and the roast chicken takeaways in Amman were a surprising treat. Further south the ancient city of Petra with its narrow defile entrance, its rock-carved treasury and monastery high above took our breath away.

In a few short weeks we learnt more about the world than we had in the previous three years.

From Jordan we crossed into Saudi Arabia and headed down to Bahrain via Riyadh. The roads were fantastic, super threads of black tar crossing an empty quarter. In retrospect it was quite an accomplishment or turn of luck, for even today it's still pretty impossible to get a Saudi visa. Perhaps the mention of our college friend whose family lived in Bahrain helped.

Petra, Jordan 1974

The following extracts are taken from letters sent back to Chichester from various points along the way:

13 July 1974: letter from Bahrain

Abdul's House roof, Gafool Road, Manamah
Crossed into Saudi, no hassles. 110°F and I've got a cold, can you believe it. Anyway we took a wrong turn somewhere after the border but eventually got it right. Passed through some incredible scenery, a bit like Petra, with massive sandstone cliffs of red rose colours. Made it to Tabuk, a

larger town, then on to Medina. Being infidels we had to take the ring road; luckily there was a fuel station.

7 July
Happy birthday I keep thinking, stuck in the back of the Land Rover flat out with terrible stomach cramp and a cold. Hugh has sunstroke and is feeling bad. Pete has been driving a lot. His visa was issued for only 3 days, while we had 5 days, so we have to get to Riyadh to extend that in 3 days. Riyadh is modern; there aren't any old places. We couldn't find the visa place so we called into Marconi's office on the street and someone helped us to find the right building. No bother to extend the visas, so we were away quickly to Dhahran.

On the road to Tabuk in Saudi Arabia 1974

Passed some amazing gas & oil flares on the way; didn't need a flash to take a picture even though it was dark. We had to fly from Dhahran across to Bahrain, about 15 minutes' flight; we left the Land Rover at customs. Had some great food and entertainment at the BAC compound nearby. There are about 700 expats living in Dhahran in various compounds. In Bahrain, Abdul was still away but his father came to meet us, dressed in his sparkling, long, white Djallaba thing and resplendent with red and white headgear. He whisked us off to his house for dhal soup and salads. He insisted we stay in the nearby Seef Hotel, which was great as it had AC.

Mr Alawi senior took us around the island next day, visiting his old house, the spice souk and eating a family meal all sat around a large 3ft-wide silver bowl. The food was great. Also went to the Awali Club and played ten-pin bowls. Later there was a dance, but it was all Indian and Pakistani girls so that precluded any participation. Mr Alawi helped us to get new Saudi visas, which was a big relief as we were certainly pushing our luck a bit leaving the vehicle in Dhahran without a re-entry transit visa. All was well; we spent a relaxing day at the house.

19 July: letter from Shiraz

Flew back to Dhahran. Hugh sat next to the pilot as the plane was overbooked. Sorted out the customs easily and found a note on the windscreen inviting us to visit some people at the Aramco complex. John, his wife, Trevor and a girlfriend offered us some drinks. A brew concocted in the complex. They had been very curious to know how an English-registered vehicle came to be parked at Dhahran airport. John's wife used to live in Bognor Regis. We got to sleep in a portacabin with AC. This diversion across Saudi has been so interesting, if a little too hot.

Kuwait City
Pete and a local Kuwaiti driving at high speed chose the same path at a roundabout and just touched; luckily the fellow kept driving so we shot up a side street out of the way. No real damage; a quick tap with a hammer sorted it.

We are looking for shipping between Kuwait and Iran, as the Iraqis will not give us a visa even for the 3 hours transit by road. Sweltering hot, tired out just walking about. Went to the harbour and met a dhow captain, Unis Mohamed. These dhows are made of thick wood, about 90ft long with cabins at the back and diesel engines and sails. Stayed at the youth hostel, where we met Indigo Overland again. Together with Indigo, we managed to arrange the dhow through Unis Mohamed for £130; we paid £30 for the Land Rover.

Had a problem with the hostel manager. We were told to eat around the back and not to eat in the same place as the girls from the Indigo trip, the separation thing taken a bit far. Don't like this place much. So much money around but all the work is done by immigrant Indians.

16 July
Been down here in the docks for hours doing the customs and passport formalities. Now the tide is out so we still can't get going. They lifted the two vehicles on by crane; it was very dodgy-looking.

17 July
Finally got away at dawn, having slept on the deck in the harbour. The captain entertained us for a while with his jokes in pidgin English. Sailed all day until 5pm and then spent 2hrs at some Iranian checkpoint. Set off up the Euphrates with Iraq on one side and Iran on the other. It was a beautiful sunset, sitting on the cabin roof, calm waters, palm trees and red sky. Arrived at Khoramshah about 11pm, so we slept on board again.

19 July
Took most of the day to clear customs, with some bother over landing fees and other miscellaneous charges. Drove on into the Zagros, crossing a pass over 8000ft high. On to Shiraz, a pleasant cooler place with tree-lined streets and lots of blue domes. Stunning actually with the interiors full of mirrors and glitter. We never seem to have enough money but change US$20 at a garage for fuel.

Stopped off in Esfahan and couldn't believe it was real. The Shah's Mosque, Friday Mosque and crazy bazaars kept us enthralled all day. Sat by the Khaju Bridge a while and later watched all the smokers with their hookah pipes.

Loading the Land Rover in Kuwait

Imam's Mosque, Esfahan, Iran

Typical bazaar scene in Iran

Kabul, Afghanistan
Had a good run from Tehran, crossed the mountains past Mt Damavand to the Caspian. Hot and sticky again. Then in Mashad we visited the blue mosque complex with its teeming bazaar around the holy shrine. Took ages to get into Afghanistan; some shifty characters at the border. Herat was fabulous: mud city, ancient minarets and a great intricately decorated mosque. Lots of horse-drawn carriages, turbaned men and women in black, totally covered up. Even their faces are covered with a sort of mosquito net across the eyes.

Herat bazaar, Afghanistan 1974

Had some problems with the fuel pump again, stuck for hours between Herat and Kandahar on the desert road. A bit scary actually. Managed to get to Kandahar. Camped at the Mayfair Hotel. Bought a milk churn. A local welding shop put a drainpipe in the bottom and we connected it directly to the carburettor. The churn is sitting in the middle of the spare wheel on the bonnet, which is OK but blocks the view a bit. It's been working well up to Kabul. Passed Ghazni, a once-famous town, where a motorcyclist bombed past, turned around and came over to chat. He was Polish and going around the world in 80 days. Since he's already been going 30 days across North Africa and Asia, he seems to be a bit behind schedule. No matter, off he bombed. I had another dose of the bug again.

We spent most of our time relaxing here in Kabul along Chicken Street, where there are loads of pie shops selling strawberry pies. Lots of hippy types and crummy hotels. The locals are great here, so impressively dressed; they spend all day drinking tea from small glasses and pouring the stuff from small, dirty teapots. The local bread is great, flat and good when it's still hot. The occasional stone is extra. We're camped in the back of a hotel. Been on an excursion north to the Salang Pass, but didn't get to the Bamiyan Buddhas as we had more fuel problems. That's a pity, but it's a bit remote out there. Met some of the Indigo crowd again at the Metropole Hotel in Kabul. The driver seemed down in spirits. Off to the Khyber Pass tomorrow.

3 August: letter from Lahore

Roof of YMCA, Lahore
Crikey, I thought the Middle East was hot, but we didn't have all the sweaty stuff. Of course it's the monsoon here. The Khyber was something else. Lots of forts, quite a good road, the place at the top, Landi Kotal, was like something out of Kim, devious-looking locals with guns but quite friendly. There were various army brigade signs painted on a rock face and then a twisting road down a sinister-looking valley.

Landi Kotal brigade signs, Pakistan

A taxi followed us down the pass, people packed inside and hanging on to the roof rack without a care for falling off. The driver was looking at us most of the time and kept getting very close or miscalculating his speed on the corners. As soon as we got to Fort Jamrud with its arch across the road we started feeling the incredible humidity.

On the Khyber Pass 1974

Visited Peshawar market, another chaotic place smelling of spices and petrol from badly smoking three-wheelers. Great people in turbans, baggy suits and trousers. They all seem to be shouting at each other all the time. There are guns for sale everywhere. The GT Road was cluttered with all manner of vehicles, people, donkeys and bicycles, with only one road rule – the biggest wins. We decided to camp a bit out of town down a sidetrack off the GT Road.

Not long after stopping Hugh dashed into the bushes, a call of nature again. Seconds later some locals arrived, certainly curious and certainly well-armed. It all got a bit scary. Pete's watch was stolen in the confusion while Hugh squatted it out well hidden from view, a nervous spectator. In the event nothing further happened but we decided to move on after dinner and ended up camped by the river Indus below the famous Attock Bridge. A great spot.

It poured all night and the road was badly flooded in Lahore, with sewage floating about all over the shop. The population of Lahore seemed to be outnumbered by enormous buffaloes everywhere, even in the Mall Road where we slept on the roof of the YMCA. The huge Badshahi Mosque was a magnificent sight, particularly with great reflections in the field-sized puddles of the main courtyard.

14 August: letter from Delhi

5 August: Amritsar
Conked out in a huge flood area, leaving Lahore with wet electrics, but some boisterous helpful local came to our aid, pushing us to dry land. Drove to Amritsar and headed for the Golden Temple down increasingly narrower streets. The streets were very clogged up, mostly with 3-wheeled cycle rickshaws pedalled by thin men with even thinner legs. We had to put on headscarves and take off our shoes. The temple is incredible; gold leaf covers the holy part and it's set in a small lake. The Sikhs are great and very impressive in their colourful turbans and gowns. In the central temple priests are reading the holy book Grant Sahib; they read for days, there's a 6-month waiting list and musicians play most of the time.

11 August: Dal Lake, Srinagar
Fuel costs 13 pence a gallon here. Had to wait hours at a check-post near Jammu. Hugh and Pete ill again, must have been the meal in Amritsar. Pretty spectacular road up from Jammu to Srinagar, winding and very slow, took us nearly 3 days. It's still cloudy and humid but some sunshine too.

Visited Gulmarg, a Swiss-looking place in the Pir Panjal range. The houses are wooden and very quaint. It's still snow-covered higher up and they have skiing in winter. I can't imagine that in India. The lift looks pretty ropy and sags between pylons a lot. We drove up to Kilenmarg with a great view over towards Nanga Parbat, Nun and Kun etc.

We are on a houseboat called New Year; it's next to one called Cherrystone on Dal Lake. Indigo are here too. Had dinner on the boat: duck, roast spuds, veg and dessert of walnuts and apricots. And tea, of course. It's all very sedate, with sellers constantly calling by selling stuff, papier-mâché, sweeties, macaroons, carpets, hash, suitings and dyeings. Jeff from Indigo played his guitar in the evening while Dave, Trish, Loraine, Paul and Phil solved the problems of the world. The hookah pipe appeared later and problems were being solved more easily.

Last night our boat starting sinking and I had to visit the loo a few times while the boat was bailed out all morning. After the sweeties man had dropped by to sell us chocolates, we climbed the hill above the lake for a fabulous view – a water paradise stretched ahead, with willows and gardens and rows of houseboats.

Dal Lake panorama, Srinagar, Kashmir, India 1974

14 August: New Delhi campsite, Asif Ali Road

We are camped between two of Delhi's main roads; it's the official camping spot. The noise is deafening. Impossible to sleep, it goes on all night. We have discovered that we should always ask a thin taxi driver to take us around, because the fat ones always charge more. Visited the Red Fort, Jami mosque and old Delhi. It's a smelly place, it keeps raining and the place is always flooded. The people are amazingly calm amongst it all, but do get on your nerves a bit. Always asking questions. I don't know how many times they asked, 'What is your country?' Couldn't get a permit for Darjeeling, so might not bother with that.

20 August: letter from Benares

I don't think I want to visit India again in the monsoon, it's so humid and mosquitoes have been tearing us to bits. I clobbered a holy cow on the road to Jaipur, when a truck cut off our space and I had to swerve on to the verge. We didn't stop to see the damage; it was still standing as we disappeared. The locals weren't too pleased, though.

Elephants at Amber Palace, Jaipur, India

There were lots of camels all along the road to Jaipur. The pink city is pretty interesting, with painted shops, palaces, and of course we saw the Palace of the Winds. It's actually quite a thin building, mostly façade and not much else. It's yellow on the back as well. The Taj was incredible; the tomb inside wasn't much but the smell from the joss sticks was very pleasant. We camped in the garden grounds near the main gate. No one bothered us, surprisingly. Sunset was great.

The road from Agra to Allahabad was diabolical, potholed and narrow. Allahabad is completely dreadful, total chaos, narrow roads, no signposts. We got hopelessly lost, going deeper and deeper into narrower and narrower, dark alleys. Spent hours trying to find a way out of this bazaar area, like some ghastly maze. Took a boat trip along the Ganges in Benares; the river was in flood and the boatmen had a hell of a job. There weren't many bodies being burnt. Afterwards we got taken to the Golden Temple – not much to see except dirty monkeys. Then we ended up in a silk shop in the backstreets. The place smells like a cowshed with cow dung everywhere, the houses are falling to bits and people smoke funny-smelling tobacco leaf roll-ups called beedies. There are millions of rickshaws, not many vehicles. It's impossible to drive around here.

29 August: letter from Kathmandu

Pokhara
Had a puncture just before the Nepalese border, seemed like we were never going to leave India. The road was even more diabolical. It's very poor here; we got lost in a place called Gorakhpur. What a dump.

What a relief to get into the hills again; super drive up through Tansen to Pokhara. After India this is really paradise, even though it's the wrong time to arrive here; it's raining on and off and very sweaty. Pokhara is a tiny place. There are a couple of pagoda-style temples and one long main street with toy houses and chickens in the street. It's like a village of farmyards. The houses all have thatched roofs and reddish-coloured mud walls. The lake is further away, pretty quiet, not much happening. There's a place called the Swiss Café, but it's nothing like the real thing, loads cheaper though. The food looked dubious and the tea was awful sickly sweet stuff – that's all we had.

Camping at Withes Hotel, Kathmandu 1974

23 August: Kathmandu
Left Pokhara having seen a bit of the Fishtail mountain, Machhapuchhre – the top we think, not much else. The road was good to Kathmandu, single lane but very pretty; the last big hill had the Land Rover struggling, though. Then we crossed a small pass and the road became a muddy quagmire. Seemed to take forever to get into the town; we passed lots of quaint farmhouses and other houses along the road. They all have sloping roofs with wooden struts holding up the overhangs. The kids just play on the roadside in all the mud. Crossed a tumbledown-looking old bailey bridge and came into the outer part of town.

We parked up in a hotel garden before the stadium, a new-looking ugly concrete and brick place. The hotel is called Withes – I don't know what it means.

25 August: Kathmandu
Just spent the day in the old city area around Durbar Square. We had some Swiss rosti and buffalo steak at the Swiss restaurant. It was great; now waiting to see what the stomach thinks of it. There are quite a few hippies around; they are hanging out in a street called Freak Street. It's full of tumbledown houses, cafés and equally tumbledown hotels. There's a place called Eat at Joe's, which has buff steak and chips for 6 rupees, that's about 15 pence. To get to the square we have to walk for about 15 minutes along these narrow streets with people spitting out of first-floor windows. They don't have glass, just wooden planks or small shutters. Everyone seems to be coughing. We passed some of the pagodas – three-tiered structures with wide platforms and steep steps up.

The kids are following us about but are no bother; some clever ones speak quite good English. Just behind the square is a crazy street called Pig Alley; it's certainly full of pigs down near the river. It's also full of low-ceilinged tumbledown houses with little cafés in the street-level rooms. One is called Pancho's, another the Upper Crust Pie Shop, and another is Chai & Pie. The street is filthy; I can see why people are always sick here but the apple pies are great. They also have chocolate cakes, brownies and other cakes of ill repute. We keep banging our heads on the low doorways.

Old Kathmandu 1974

29 August: Kathmandu
Been here for days but it's all very relaxing, up late, stroll to Durbar Square, mid-morning chai. Lunch at Joe's, more chips and buff steak, then it's afternoon pie at the Upper Crust. Back to the hotel. Last night we walked into the square and found some musicians playing, a squeeze-box and one with a flute thing. It was great. Some hippies sat nearby smoking away. Some of them are in a bad way; one Italian guy can be found in Freak Street lying in the gutter half the day out of his head. He begs from locals and won't last long.

Visited Patan, another small place across the river. It's got many pagoda temples and a square. There's grass growing on each tier and they all look about to collapse. One temple has bamboo scaffolding around it and the grass was being pulled off, so I guess they are doing some restoration work. It's King Birendra's coronation in February, so the place will be all cleaned up.

Visited Pashupatinath – the burning ghats – the river is filthy, and Swayambhunath, the monkey temple on a hill overlooking the valley. Also went to Boudhanath, a gigantic round stupa where Tibetans, Sherpas and other Buddhist people circle around the stupa clockwise turning hundreds of prayer wheels. It's a great place with all-seeing eyes. The stupa is surrounded by quaint brick and wooden houses. There's a small monastery on the west side and beyond that it's all fields and quaint farmhouses. It's about four miles from Durbar Square through a place called Dilli Bazaar. The houses are all brick and wood in Dilli Bazaar as well. It's like Toy Town; the doors are very low. Cows walk about in the streets, there aren't many cars. It's like a country village but full of dirt, heaps of dung, piles of rotten weeds and half-clothed kids. It's pretty hot still and rains some of the day. The place is very green, though, with plants growing everywhere, even out of buildings.

Had lunch at Joe's with Hugh, Pete, Laurie, Phil, Maggie and Paul from Indigo. They're all going their different ways now. Had dinner at the Tibet Dragon restaurant with a South African girl. Met the Indigo crowd later at the Travellers Lodge.

It looks like we are going to leave the vehicle here in this hotel for a few months, as shipping from Madras to Australia is too expensive. Hugh and I will come back in February and go on trek to Everest. It will be sad to leave. Pete has decided to stay here a while and go on trek in a couple of weeks after the monsoon. The Indigo driver has quit and is going to a monastery to find his way.

Patan Durbar Square 1974

Old Kathmandu panorama 1978 (Photo taken by Bob's brother-in-law Phil Mitchell)

9 September: letter from Bangkok

Had a good time in Burma. Very interesting and we didn't get sick. Flew into Rangoon and stayed at the YMCA. Unfortunately my whisky bottle was broken in the bus crush from the airport, so I couldn't sell it! Rangoon is terribly rundown; the buildings haven't been painted since the Raj left.

Main street, Rangoon, Burma 1974

The Swaydogon pagoda is out of this world. Gold leaf with unbelievably decorative temples, shrines and pagodas all around the central pile. The only hotel is called the Strand, a grand rundown heap. The people are very mellow and very friendly. We visited Pegu to see a large reclining Buddha and then Mandalay and Pagan in our week here. Had to sleep on tables at the YMCA in Mandalay under our new mozzie nets. The whole town was surrounded by vast floods of brown water. The massif Mandalay Palace stands on a hill with panoramic views of the watery plains.

Swaydogon Temple, Rangoon 1974

Pagan temples, Burma 1974

Pagan was incredible, with loads of massive temples dotted about the plains. Some were falling down but it's a peaceful place all right. The transport from Pagan was on trucks and minibuses that kept stopping or breaking down. We waited hours at a place called Kyaukpadaung for the night train back to Rangoon. Flew on to Bangkok.

> **Myanmar Today**
> Almost since my visit in 1974, Burma as it was known has remained under a military dictatorship. It has nearly always been possible to visit, but only for a week or two. The release of Aung San Suu Kyi was the first tentative step towards change, and more have followed. The buildings seem to have been painted and the Strand Hotel has been upgraded. Apparently it is no longer possible to fund a week in Myanmar on the sale of one bottle of whisky!
>
> How will the country fare under the new democratic system?

16 September: letter from Bangkok

11 September: Vientiane
Just waiting here to wangle a ticket on Royal Air Laos to Luang Prabang and on to Ban Houei Sai. Anyway got your letters from the Bangkok Poste Restante; the card from Herat and letter from Lahore must be lost.

Got my Lao visa OK. Hugh decided not to come to Laos, so I will see him in Malaya or Aussie. The Thai trains are great after Burma. Overnight to the border of Nong Khai. Crossed the Mekong on a small boat. Vientiane seems to be asleep all day. There are a few cycles but virtually no cars. Some cranky old buses. The people are again very friendly, though. I visited the That Luang, a temple complex, quite different from the Thai Royal Palace in Bangkok. It's blackened by weather. My room cost 40 pence and they sell American cartons of milk here, which is better than sterilised water all the time.

Now I know why the room was cheap; there's a noisy disco under my room. Went to have a look, it was full of local girls and some Thais. The disco is called 'the Green Latreen'. Latrine – yes, really. It turned out that you had to pay for a drink before you could actually dance. The girls are very pretty. It also turns out that they charge between £1–£2 for added services. They used the most astonishing language to describe these added extras – all in backward arranged English. Some westerners came in later; I was very tired so retreated while the going was good!

16 September: Bangkok
Glad to be back in Bangkok; the trip to Laos was incredible but a bit scary. I got to Luang Prabang, another sleepy town with temples on the hill above the Mekong. The colours are wonderful: blue hills, lush green forest and muddy brown rivers. I took a boat trip with three other travellers up the river to the Pak Ou caves.

There were hundreds of Buddha statues in the caves, the only access was by boat, and the caves were under a massive overhang. The food is mostly noodles and the hotel is rundown in Luang Prabang. There is a great view from the temple on the hill near town.

Across the river the communist Pathet Lao held control. We had to go over to their checkpost with our passports. It's a very strange set-up here, with the Royal Government holding part of the country and the communists the other part. Both sides have an uneasy truce at the moment, with both using the short runway at Luang Prabang for landing their planes. It wasn't safe to travel by road.

Note: The Pathet Lao took control of the country a few months later, when the North Vietnamese captured Saigon in April 1975.

Pak Ou cave along the Mekong; Cave Buddhas

Had to fly everywhere, but it's very cheap. Went by DC3 to Ban Houei Sai then across the river back into Thailand. Couldn't find the customs or passport place anywhere. There was one place – a shed – but no signs and no one about. Took the last bus to Chiang Rai, as there was no place to stay at the border. Got back here from Chiang Mai, a small town in the north with some nice golden temples.

26 September: letter from Australia

26 September: Perth
It seems strange to be back in a modern country after the last four months. Had a pretty fast trip through Thailand. Had a problem at the border at Hat Yai with police, as I didn't have an entry stamp. A bit unpleasant; they pushed me about a bit. It was the worst treatment on the whole trip. Probably because there are problems with Laos and also there are a lot of doped-up travellers about. Southern Thailand also has some insurgents. Things were bad until the big boss came in. He was good, spoke English and sent me on my way the next day. They could see the Lao exit stamp anyway. Whizzed down to Penang and Kuala Lumpur. Then visited Malacca, a nice place with lots of red-painted buildings. Singapore seemed quite modern except for the old area around Bugis Street. Haven't seen Hugh anywhere en route. Decided to fly to Perth, didn't fancy being seasick.

Well, the plane was late, British Airways too. Still, loads of English newspapers to read. The film was called The Dove, about a journey around the world, seemed fitting. Had a great view of the lights of Perth as we flew in. Can't get used to being in a 'normal' place.

****

Hugh turned up a few days later, having taken a Russian ship from Singapore to Perth. We both stayed in Aussie for three months, working for two months and then travelling east to Sydney by car. In Perth I stayed with a college friend, George Gilchrist. I worked at the Royal Perth hospital as a porter, the same job as in Chichester before the trip – I was well qualified! My sister flew into Perth to join us and later went on to New Zealand. Our plan had been to go to Alice Springs and Darwin, where we would sell the car at a profit before flying to Bali. However, Cyclone Tracy devastated the city of Darwin on Christmas Day 1974, putting an end to our trip north, so we had to drive to Broken Hill, sell the car at a loss, hitch to Sydney and fly from there. We spent a few days in Bali, mostly on the beach, got sunburnt under a cloudy sky and then headed west.

Back on the road: January 1975

Returning to the diary and letters…

16 January: Train to Jogjakarta
Alas my t-shirt has finally given way, and my other one will only just make Kathmandu I fear. Out of the window we can see Mount Merpati volcano and endless green flooded rice fields. Hugh is laid out and a smart Javanese lady is sitting nearby in a very colourful sarong-style gown. It's boiling hot as usual and my sunburn is really bad news. Bali was full of Aussies. We had a room in a house for 30 pence including breakfast. The train has just stopped and we have been presented with some flowers thrown through the window at us by two nice girls. On Bali we hired some motorbikes at Kuta Beach and explored the interior. Visited the artisans at Ubud and a lake further north. Great fun. Crossed on a ferry from Gillimunuk to Java and stayed in a place called Bangyiwangi. We are going to visit Borobudur and the Prambanan Hindu temple in Jogja and then go by bus to Jakarta. Probably have to fly to Sumatra, to Padang.

30 January 1975: Colombo, Ceylon
Had a pretty good trip through Sumatra. The bus from Bukittingi to Lake Toba was horrendous. We broke down overnight and ended up sleeping in a village in a straw-roofed house with loads of bugs and rats. Sibolga was a strange place, with a colourful mosque and not much else. Been eating peanuts, pineapples and bread. I don't like the cold Indonesian food much. Lake Toba was great, though, and full of travellers.

Flew from Medan to Penang then via Kuala Lumpur on to Colombo. Ceylon was a lot like India, but a bit slower. I quite liked it; after the hassles of India in the monsoon this was okay. The beaches are great south of Colombo; not many travellers or tourists about anywhere.

Badly dressed at Borobudur, Indonesia 1975

Central Kandy, Sri Lanka 1975

We took a bus into the hill country where tea is grown. Some drunks on the bus ended up being booted off, but not before a fight broke out. Kandy was cooler, with the pink-coloured temple of the tooth to see. We also went to Sigiyria – a huge, almost inaccessible rock with some frescoes painted on the rock faces. The stupas at Anaradhapur were like the ones in Nepal and the sleeping Vishnu at Polonurawa was quite spectacular.

The ferry from Talimanar was hell, mostly passport control hassles. We got to Rameswaram in India. What a place! It was full of temples, monkeys, filthy squalor, dark alleys and thin, dark people. Actually it was a bit like Benares old city, but hotter. The noise outside the hotel was incredible. Took a train to Madras for curry.

11 February: train to Calcutta
We met a couple of Mormons dressed in suits in Sydney. They insisted on giving us the book of Mormon. This may have been lucky, depending on your point of view, as it was the only book we had on the 48-hour trip up to Calcutta. What a trip! The train never seemed to get up to speed; we spent hours in stations, rocked about at night. Food was only available at the stations, so we starved most of the way. The book is being read from cover to cover by both of us, but that's all you can say about it really. Should be in Calcutta tomorrow morning, God willing. It will be great to get back to Nepal. I wonder if the Land Rover is still there.

14 February: Kathmandu
Took another train to Patna. What a dump! Flew to Kathmandu in a Fokker. Had a great view of the Himalaya coming into the airport; we could easily see Everest, Gauri Shankar and Langtang. The air was so clear, no monsoon clouds this time. The valley was draped in the colours of spring. The Land Rover was completely untouched, safe and sound. It was great to be back, like a second homecoming. We talked a lot to the hotel manager. Even managed to get trek permits and flight sorted for Lukla – $35. Had lunch at Joe's and ate all afternoon in different pie shops. Tried Aunt Jane's chocolate cake – wow! Moved on to the Yellow Star and then the Tibetan Dragon.

17 February: Kathmandu
We seem to be eating constantly these days. Found some great lemon meringue pies at the 'Chai and Pie' this time. The steak and mashed potatoes at the Yin Yang was great, a bit expensive though at 12 rupees – a dollar. Got some great photos from Swayambhu this time: clear skies with mountains on the northern rim of the valley. It's a nice walk out from the city, quite a long way across the river and through leafy country lanes. Hugh had four pieces of lemon pie yesterday at the Chai and Pie.

Coronation rehearsals, Kathmandu, spring 1975

Wandered into the square yesterday to find it all blocked off. A bit later, after some more pie in Freak Street, we heard some bands playing; lots of soldiers in different colourful uniforms marched into the square. Then came about twelve elephants all dressed up in magnificent red silk cloth with heads covered in gold. The last one had a large wooden platform all in gold. The streets were crowded with onlookers, and photographers from Time and Newsweek scurried about. We joined the throng with cameras ready. This turned out to be the full dress rehearsal for the coronation of King Birendra.

June 2001

The king is dead. We are in Etwall, Derbyshire, at Siân's parents' home, and the BBC Radio 4 news at 7am is detailing the horrendous events at the Royal Palace in Kathmandu. It seems beyond comprehension. Could the son and heir to the Nepali throne really have killed his own parents, brother and sister?

Our thoughts wander back to the streets of Kathmandu. Mine flash back to the coronation rehearsal in 1975, such a happy, joyous day with all the pageantry and glitter, the brilliant red silks of the elephants, the man from Newsweek pushing by to get a photo, the raptures of the crowd, the cheeky kids and the glistening peak of Ganesh Himal in the morning light.

Siân is thinking about the day the king came from his palace to stroll down Durbar Marg, to meet his subjects with an air of informality not normally associated with the monarchy. This street party was being held to celebrate his fiftieth birthday. Following the introduction of the new fledgling democracy, the king had become even more popular. He passed by as we sat with KC at his cake stall, pausing to see the activities.

It is the passing of an era, the death of the old Nepal and the birth of a new, somewhat unpredictable, era. Again our thoughts focus on Kathmandu, the spring of 1990 when we were working as leaders for Sherpa Expeditions in Nepal. That was a very different story, one that we will tell later. That was the birth of democracy in Nepal.

Fifteen years earlier, Hugh and I were on trek in the Himalaya for the first time...

19 February: Namche Bazaar
Resting up in the Himalaya 'Hotel' in Namche at 11,000ft.

Flew to Lukla yesterday on a Twin Otter plane, about 15 seats. Departed at 7.45am and had a fantastic flight, going low over ridge after ridge, with hillsides ablaze with rhododendrons. The view was stunning, crystal clear sky. As we came into land, we could see the airstrip through the pilot's window. It was hairy stuff, the landing. The mountains were dazzlingly clear. Got to a village called Jorsale in about five hours and had lunch – rice with dal, more rice and more rice, that's all. Had to cross a nasty bridge with no sides and only logs stretched out on logs. The river was very fast-flowing below. We've seen some low-altitude yaks called dzopkios or something like that.

Left at about 3.30pm to climb the big hill to Namche; didn't have any maps so didn't know how far it was. Got steadily slower and slower going up the hill; soon it was dark. Eventually saw some lights and after another half an hour we arrived in Namche knackered. A monk stood in the moonlight above a nearby terrace, just past a small stupa. He came over and helped us to find the lodge – the Himalaya – in a corner of the village below the monastery. It's a tiny place with just a few houses and this teashop lodge. Everyone sleeps in the same room. We were fed omelette, vegetables and pancake. The room was freezing at night, so we didn't get much sleep and had bad headaches.

Today we took a stroll up to the ridge above the village and had the most fantastic view. We could see Everest above a massive rock wall and on the right Ama Dablam, the most fantastic peak you could ever see. A plane came in, missed three peaks and landed below me on the airstrip above Namche. It soon took off again. There were some celebrations in the village for the king's coronation in the afternoon. After that I walked around the Namche bowl above the monastery to the Thami valley for a view north towards Tibet.

Note: Lukla to Namche in one day should not be on anyone's agenda today, knowing more about altitude sickness!

Namche Bazaar 1975

'Lodge' in Deboche 1975

20 February: teahouse below Thyangboche
It's snowing outside now and flaming cold in here. The fire smoke is a killer on the eyes. We walked for about four hours to the famous Thyangboche monastery. It was a big climb up the hill from the village at the bottom called Funky Tengy. There were some waterwheels driving prayer wheels there. Just polished off spud, our own sardines and some tsampa – a dry chickenfeed mass of roasted barley and tea mixed together. Sounds awful. This lodge is only built from stones and wood. It's freezing. We have to share the floor with the locals. It's a mud floor and a bit wet from the leaking roof. Three Sherpa girls are running this place.

21 February
Didn't sleep much; so cold, melted snow kept dripping on to my face. The bag is okay but it's the floor that's so cold. Our thin mats are not much use. Slept in all my clothes. Now having breakfast, it all takes time so plenty of resting. Had an omelette, now waiting for chapatti and tea. There is about two feet of snow outside. The Japanese Women's Everest expedition is due here in a couple of days.

22 February: Pheriche
Alas, yesterday it snowed all day, and we walked for hours, slipping about. Saw some real yaks at last. Trekked to Pangboche for lunch of biscuits and tea. Then another three hours to reach Pheriche. Today there are high clouds. We have a superb view of Tawache peak, with Ama Dablam behind and Lhotse ahead above the ridge. Didn't sleep much again, so cold and got a neck ache today. Here comes breakfast. Rice pudding.

Lobuche
Now at the height of 16,100ft of misery, according to our handout from the Kathmandu tourist info place. The trek up was sheer agony, but the views were incredible. Snowed again later, though. Got a bad headache tonight, Hugh is also feeling bad. Again we are lying on the floor to sleep and again the roof is leaking badly. Didn't feel like any food. I suppose this is what they call altitude sickness.

Next day. No sleep; splitting headache too. Six others have gone down. Hugh too ill to move. I went on with the Austrians. It was partly clear, so we had some good views. Saw Pumori clearly later on. The snow was over two feet deep and soft, so we made terrible progress. Climbed a steep rocky area and could see Kala Pattar but no hope of making it to the base of it. Took about six hours to get about two thirds of the way. Felt terrible, head splitting, stomach groaning and a bit dizzy. We all decided to go back down to Lobuche. No sleep.

25 February: Thyangboche
At least I can remember the date today! We all staggered back down as fast as we could yesterday. The day was much better. Poor Hugh had a terrible trip down, feeling very ill. He always got to the places ahead of me. A raven has just whipped my biscuit away – damn. Pretty awful taste anyway. It's superb today, with clear blue skies. It's great sitting here by the monastery with mountains all round. They look so pristine in fresh snow. Excuse me while I go and film the monastery… no it's not right yet, too much shadow. Everest looks great from here.

26 February: Namche
Just got up; it's a bit warmer now. A Tibetan Lama has been mumbling prayers all night. One Tibetan girl here speaks some English and has been telling us about Tibet. Pity we can't go there. Had chapatti, eggs and tea for breakfast. Looking forward to apple pie at Pancho's.

Later. Made it down the hill okay and met the porters from the Japanese expedition, twelve ladies heading for Everest. Hundreds of porters carrying boxes, even ladders. Back at Lukla now completely knackered. Smoke gets in your eyes.

Note: Junko Tabei became the first woman to summit Everest from the above expedition.

27 February: waiting at Lukla
No planes yet, rained again last night. The radio here is out of action. Really noisy night, with kids bawling, dogs barking incessantly and snoring locals. Well, it's 1pm and no planes, so that's that. Apparently some people have been waiting five days so far. Had some pineapple juice, a product of India. Hugh lost 105 rupees somewhere and has decided to walk to Phaphlu down the valley, a couple of days' walk. I'm planning to stay here longer. Some people are getting angry here. It's a bit frustrating, but maybe tomorrow…

Flying over Boudhanath from Everest 1975

28 February: Kathmandu

Well, they sent in a few planes and we all got out. Fabulous views again. Poor Hugh should have waited. He seems to be lower on money. More buff steak, chips and pie. Hugh turned up a couple of days later, having had a gruelling three-day walk to Phaphlu over countless ridges. He met Edmund Hillary, though, at his school and flew back the next day. We prepared the Land Rover and sadly departed Kathmandu after just a few more pies and cakes.

Kathmandu Valley 1975

21 March: letter from Tehran

There was a fuel shortage in the Indian state of UP across the border, but after that we made good progress to Pakistan. A spring had to be replaced in Dera Ghazi Khan and the road to Sukkur was terrible. The climb up the Bolan Pass was amazing and Quetta was cool and pleasant. We crossed the border into Afghanistan at Spin Boldak, and filled up with fuel in Kandahar; a fatal mistake, as the fuel must have been adulterated.

En route to Herat the engine suddenly seized and we coasted downhill some 20 miles south of Herat. We later learnt that Afghans and others in the subcontinent are in the habit of mixing fuel with other dubious liquids and that such a mix very likely caused the engine failure.

We spent a week in Herat but, running out of money, we decided to abandon the vehicle. Customs were none too pleased with us, as we had sold a lot of items from the vehicle, leaving it a mere skeleton of its former self. Eventually we had to take it to the border, which meant an expensive tow job. The Afghans were pretty difficult, but the Iranians allowed us to donate the vehicle to the customs in Taybad, so after a lot of paperwork we said a sad farewell to the Land Rover, long since named Calamity.

As it all was.

Kandahar, Afghanistan, spring 1975

Towed to the Afghanistan–Iran border, Hugh 1975

April 1975: back in the UK

We arrived back in England in April, somewhat wiser, poorer, but full of the joys of the pies of Kathmandu. The hardest thing about the whole trip was getting back to reality. This meant no more days on the road, no more sense of excitement about what would be on the next horizon, no more strange lands to explore. Of course there were no more bouts of stomach ache, no more difficult officials and no more discomforts.

I was back to mucking out the pigs and collecting the eggs in the chicken sheds, just like old times when I was a kid and 'child' labour issues weren't discussed. I was grateful to my parents for their sense of duty and fair play. They weren't doing very well financially, as the farm was too small. Later when Britain joined the European 'Common Market' things got worse and they both had to find supplementary jobs.

I returned to my old job as a hospital porter. In those days it was easy to find a job. Soon I was back in the swing of the place. Had I really been half way across the world and back while the same porters had continued their usual work routine? I didn't really feel much different; shouldn't I be a different person by now? Didn't the trip do anything for me? I was still the quiet, conscientious hospital porter I'd been before I'd left.

The doctors still acted with a certain air of importance, barely noticing the porters. The nurses were still hard to encourage out on a date, although I did have a bit more success than before. It was hardly due to the fact that I was now a 'big explorer' – well at least I should have been. But these things don't matter in the real world. 'Hi, how was your trip,' some would say, barely listening to the answer. A few people showed some interest, but in general everyone had their own life to race on through.

I determined to save money as quickly as I could; after all there was a whole new continent to see. It would not be in a Land Rover this time, unfortunately. At least I was able to live very cheaply at home, a fact for which I owe my parents a great debt. I would surely never have been able to save money for the next big adventure so quickly otherwise, especially doing a fairly poorly paid job. Within six months I was back on the road.

Trans-African Ordeals: 1975–76

Once again I am reverting to the letters I sent back home. It's November 1975 and I am riding in a pink, four-wheel-drive, ex-Army Bedford truck with a bulbous bonnet and a bunch of mixed nationality travellers. We left from London Victoria, full of anticipation, nervous excitement and some apprehension, no doubt.

10 November: 9 miles from Rouen, France
We are bogged down here in deepest rural France, barely a hundred miles from London, with a damaged differential. A lot of the oil leaked out and that was that. It's a public holiday tomorrow so here we stay, winter in northern France. They are mostly Aussies and New Zealanders on board the trip, with a Yank, a Canadian, two English girls and me – the sole little Englander bloke. The drivers are British. After swapping the front diff to the rear, we finally got going.

17 November: letter from Madrid, Spain
Since I posted the last letter we have made progress in cooking better evening meals, but as far as mileage is concerned, gains amount to nil miles. We have been stuck here almost a week already; is this Trans Africa or Trans Europe? It's freezing here and we are waiting for more differential parts. We think Franco has died.

10 December: letter from Tamanrasset, Algeria
Just a line from Tam, the dead centre of the Sahara. Received your letters in Tamanrasset. This is a great one-street town, but we don't look forward to formalities much tomorrow. Surprisingly we crossed Morocco without any problems. Went to Fez and Marrakesh and then came east to the border at Figuig. Went east to Laghouat and then south into the real desert. Today we stopped in Ghardaia, a fantastic place with dark low-arched alleys and an incredible square and mosque. Took some great pictures.

Well, we broke down near El Golea; it was the fuel lift pump. Visited the date palms in El Golea and next day stopped for mint tea in In Salah, another great mud-walled town. Hit the piste after that and had a great camp near Tadjemout, then through the spectacular Arak Gorge and into amazing sandy plains with gigantic rock outcrops hundreds of feet high. Came around the western side of the Hoggar Mountains.

Windblown 'Fuzzy Wuzzy' Bob at Assekrem, Hoggar, Algeria 1975

En route to Tegguidan Tessoum, Niger 1975

14 December: Agadez, Niger
All is well, only 10 days shopping time to Christmas and here we are on the bottom side of the Sahara in this fantastically old and colourful town in Niger. We hadn't seen another vehicle for the last two and a half days until now. I was on cook duty at the border at In Guezzam, a bit of a hellhole. We had dehydrated chicken curry and Angel Whirl. We've been covered in dust for days and everyone is getting sore eyes and throats.

The local people are so colourful. We stopped at a camel, cattle and goat watering hole near Tegguidan Tessoum. They pulled the water up hundreds of feet with an old inner tube bucket and a tatty rope pulled by camels. The truck got stuck in mud and we hardly got anywhere that day.

This place Agadez is amazing. There's a tall mud mosque held together by large wooden sticks, there are camels everywhere and men dressed in flowing robes with great turbans of cloth. The buildings are in mud and we are pursued by hordes of flyblown kids. There's an old hotel, which looks like a palace; some of the Aussies are into Bière Niger. Afterwards I found some street musicians near the Sultan's Palace; it's boiling hot still even at 8pm at night.

16 December: Zinder, Niger
Well, this is another fabulous place with a great market. Pete the Yank and I wandered around the place; we saw lots of strange stuff, yams and red round nuts, Lux soap with black girls on the packet, batteries, biscuits and terrible meat covered in flies so thick you couldn't see the meat.

17 December: Nigerian border
Between Niger and Nigeria. Been stuck here for hours, as usual for any border crossing. I can see a typical native hut from where I'm sitting; it's incredible. Yesterday we passed more and more thatched huts and villages as we got into more vegetated country. Lots of huge Baobab trees with strange shapes.

19 December: Kano, Nigeria
Well here we are in Kano. Received a letter from Christine (my sister) in Tehran on her overland trip. We are camped at the Jockey Club so it's pretty noisy. It's a bar, mostly, with some dubious characters hanging about. They're all called Mohamed, of course. Some of the crew have been very slack about water; they seem to find it macho to drink from the taps.

Christmas day, Cameroon 1975

23 December: somewhere in Cameroon
Well, it's finally happened, I had to scramble into the bush in great haste last night. Having sorted that, I twisted my ankle and it swelled up so much I didn't sleep a wink. Can hardly move, can't do my washing-up turn, so sad. Yesterday a bunch of people came back to the truck 3 hours after the designated departure time, so we were 'right pissed off', as they keep saying in their grating Aussie accents. Met some other overlanders, two orange Encounter Overland trips, 40 people all together. I'm glad I'm not in such a big crowd.

25 December: Cameroon bush south of Garoua
Well, all is forgiven and we had a great Xmas Day with our gang and the two Encounter Overland trucks, having a 'hungy' New Zealand-style, cooking all the food in the ground on hot rocks. We had Christmas pudding, with beef, sweet potatoes, stewed pumpkins and other burnt offerings. It was all cooked by our antipodean crew members, directing us lesser mortals.

1 January 1976: Bangui, Central African Republic
Big Daddy Idi Amin is in town with Mobutu Sese Seko Sese Banga etc for the 10th anniversary of Bokassa's rule. The show is a riot. Police are directing accidents, BMW motorbikes speeding down the only tarmac strip in the country to the airport. Operation Bokassa One, a local on a motorcycle, almost eliminated the entire presidential guard by wobbling his aptly-decorated cycle.

The notorious dictator Jean Bedel Bokassa, later the self-styled Emperor of Central African Republic was rumoured to have eaten children for amusement among other ghastly and grizzly happenings. We had to camp below his palace near the Rock Hotel by order of the police. Being in the centre of town, we also needed to maintain a 24hr rota for guarding the camp.

We all retreated to the Rock Hotel for the disco last night, to celebrate the New Year in sweaty, groovy style. I was picked out by the French Ambassador's daughter for a knackering dance session; it can't have been because of my stylish eveningwear or smooth-talking French language.

A comical New Year's parade filled the day in style, a style that only Africa could possibly put on. The entire CAR air force flew past, one plane with only one engine working; it barely missed the tall tropical rainforest trees down the road. Got a really close-up view of the pilot and Big Daddy Idi Amin.

3 January: Bangui
We went shopping for two weeks' supplies then crossed the Ubangui River, only to be turned back by the Zaire immigration. (The great Zaire expedition falters before it's begun.)

Now trying to get Sudan visas, but the ambassador is a very bored young fellow from Khartoum and has other ideas; he wants to wine and dine our girls before any visas are remotely possible. It's a good job the whole truck isn't loaded with females, or we'd never get out of here. And of course our CAR visa will need extending before we get the Sudan one and here we stay.

At least we managed a day trip down to the Congo border – that's another version of Zaire run from Brazzaville not Kinshasa. That's if anyone is running Zaire with Mobututu Sese Seko Sese Banga etc. partying in town with Big Daddy.

Our co-driver and one client have succumbed to hepatitis and both had to fly out to Paris.

9 January: Bangui
It looks like the Sudanese Ambassador is getting bored with our girls; not that anything underhand appears to be going on. Encounter Overland are also here now for visas, so they have become more interesting for the fellow. We might be out of here tomorrow with our visas.

Road in the Central African Republic 1975

26 January 1976: letter from Wau, Bahr al Ghazal, Southern Sudan

Dear All,

Left Bangui making slow progress due to deteriorating roads. Spent three days in Bangassou with brake trouble and officialdom delays. Drove on to Obo, taking twice as long as expected, due to poor roads, ferries not working and overgrown tracks that are called highways. Broke a back spring on the trailer. Next day a tooth came off the crown wheel on the back axle and made a neat hole in the diff cover. Also a tyre blew.

Just past Obo the back differential blew. With the driver Nick sick, Geoff, a fellow client and I set about replacing the rear diff with the front diff as we didn't have enough fuel to crawl along in low ratio to Juba and certainly not to Kenya. Unfortunately the bearings were smaller and the job took two days. Unfortunately, although not a total surprise, we stripped the second diff a day later on some gigantic rock steps in the road.

With the rest of the party setting up camp in the bush for some time, the driver, Nick, Dave and I set off for the border to look for parts in Wau, 300 miles on in the Sudan. With no traffic at all we had to walk the 40 miles to the border, mostly at night until some villagers warned us about large animals with big hungry mouths and fluffy, but not cuddly, manes. Elephants were also said to be a menace. We ate our stale bread and shared some tins of fish for two nights and a day. We spent the second night in another village, the people inviting us to share their modest but welcome little round thatched mud house. It was quite something – we were offered food and some straw matting; a more hospitable people you wouldn't find anywhere in the world. A pity about all the bugs in the night though!

From the border we managed to hitch to Wau on the top of an old Bedford truck, severely overloaded with a boisterous host of passengers all clinging to the sacks of grain over every bump of the dirt road. It was hell under the burning sun, stomach cramps threatening at any moment. (A great story to relate from the comfort of leafy England later – you know, great white ashen-faced hunter meets the locals and explores the Sudan by exotic lorry).

26 January: Wau
The driver and I are now in Wau and going to look for spares. Actually I had a drive of the truck the other day. It's near a village so the others can get food from there, and there is a stream a couple of miles walk away. Dave is running out of time and the driver Nick's sickness is worse. This letter is being posted by Dave from Khartoum.

Continuing from the diary:

27 January: Wau, Sudan
Dave went to the bank and managed to change money for another air ticket in 15 minutes with the help of the airline man. Out to the airport, dirt strip, Fokker took off with Dave and Nick for Khartoum. Nick left the truck carnet, papers and £16 of company funds. I returned to the only eating house in town and then started to look about for spare parts. Managed to find a pinion but no crown-wheel. Found a complete axle in the customs house, but they weren't going to part with that. Even found some gear-oil for sale. Met a couple of Europeans; one Swiss had been looking for fuel for days. A few haggard-looking travellers arrived on the train from Khartoum, a marathon 5-day trip. Slept in the police station courtyard with other travellers. No beds, just sand.

2 February: Tambura
Waited two days in Wau for a lift, a pretty dire state of affairs. No spares, no food, just melons; got sick again and felt weak. Travelled back to Tambura on top of truck again, sick as a dog under the hot sun. Met some German volunteer workers and got invited to their compound – shower and food. After days of no food, I lost the lot, being sick again. Five days to travel 250 miles.

3 February: back at the camp
Finally got away with the Germans and left Sudan. Took five hours to reach the stricken truck. Great to be back. Found only nine here now. Seven had left. Five had left just after us with suspected hepatitis, and two others had been taken to Juba hospital by some tourist vehicles, with suspected hepatitis as well. The camp was well set out with loos, a cold store on the evaporation principle, a summer-house with mosquito nets all round, and a shaded area of poles and leaves.

Very early the next day, Aussie Dave and Kiwi Mike set out to walk to the Obo mission some 25 miles back towards Obo. Mid-morning they arrived back with a big Unimog from the mission. They proposed to tow the truck and trailer back there. It took about 4 hours in the event to get there.

4 February: Obo Mission
Back at the mission and now what? Then we saw the abandoned Mercedes axle behind the mission shed. I rushed off to measure it up; the others seemed very sceptical and there was talk of walking out to Sudan. I approached the mission people and they agreed to let us use it and we could sell them the trailer for $20. It all seemed a bit crazy.

5 February: Obo
Had a great day – jacked up the truck and replaced the old axle with the Mercedes one, managed to sort out the differential on the Mercedes. We had to replace one bearing; luckily the truck's one was exactly the same. The mission blokes welded the truck propshaft on to the new axle – the only electricity and welding gear between Bangui and Juba. The brakes didn't really work too well and we had to put four bald Land Rover tyres on the wheels. The girls kept us all plied with great fruit salads, pineapples, papaya, bananas and lemon juice.

10 February: en route
Departed Monday from the mission, full of trepidation, but made it past our breakdown camp. Nearly came off the road where some terrible rock steps barred the route. Had to camp here overnight and use rocks the next morning to build the road up under the truck. Took six hours to go 50 yards. Continued at about 5 miles an hour. Had a puncture. Got the tyre off using a screwdriver, hammer and jack handle. Used one of the two patches on board. The truck's air-line kept flying off the valve connector. Used snip pliers and tweezers for the valves. Drove slowly all day, heart-stopping gullies and holes, sloping track, ruts 2 feet deep, hairy motoring! Stuck in a mud hole for two hours. Met some tourists going west; our notoriety has spread to Kenya along the bush telegraph – 'So you're the lot who've been stuck in the bush for 4 weeks.'

Changing the back axle, Obo Mission, CAR

Crossed the border into Sudan. The officials were most helpful; they could hardly believe their eyes with the truck a good 2 feet lower at the back than the front. Had another puncture; we seem to be on three rear wheels longer than four. Decided to ditch the heavy tailgates and then, for good measure, the seats we didn't need. Later a Land Rover came by from the German Volunteers and helped; they gave us another tube so we struggled on.

Southern Sudan 1975

More punctures near Maradi, Sudan

13 February: Maradi
Mended another puncture but, before we put the wheel on, Dave was attacked by some vicious bees and ran off at high speed down a village track. Suddenly everyone was being attacked and scattering in all directions, mostly to a stream to get under the water. Dave was a sorry sight, I was stung on one eye; we all looked terrible. They swelled up a hell of a lot. Eventually got going and drove all day slowly to Maradi. Went to see the German Mission and they gave us a load more old disused tyres and tubes, plus a valve extractor in exchange for two jerry cans.

17 February: Juba, Southern Sudan
Arrived here finally and it looks like we are finally done for. No fuel in town. Had to repair a front spring, feeling exhausted. Two passengers who had left earlier had ended up in the isolation wing of Juba hospital. Barbed wire fence and all; looks like a fate worse than death. (They later survived to reach Nairobi.)

18 February: Juba
Well, surprise, surprise, Aussie Dave has found a shady character with some black market diesel, so we all chipped in to cover the cost: £35 for 40 gallons. The company money has all gone, but at least we are off again. Only 600 miles or so to Nairobi via Lokichogio and Lodwar.

20 February: Northern Kenya
Drove 150 miles today all day and no punctures, that's 906 miles from Obo now. Fabulous scenery from Juba, the swamps were a bit scary but then we had great rocky mountain outcrops and then into desert country near the border.

There are some incredibly friendly local people here offering us fruits and other strange food items. After some sandy sections, we are into Turkana country, with locals completely starkers.

22 February: Kampala–Nairobi Road
Past Lodwar, the Kenyans seem very pleased to see us in our outrageous pink machine. Now getting into more civilised country. Saw some giraffes just after joining the main Nairobi road. Haven't had any more punctures.

Turkana country, Northern Kenya

23 February: Nakuru
Drove up to the Menengai crater above the Rift Valley, then into Nakuru. Had three bars of chocolate fudge and felt ill.

24 February: Nairobi, Kenya
Finally pulled into the big city, surprised that no policeman bothered us about our sloping truck. Ate almost continuously for three days: mostly bacon and eggs at Brunner's Hotel and cake at the Thorn Tree Café. Spent a great weekend at Embu with Sarah's family, who still live in Kenya on the slopes of Mt Kenya. Ate so much food! Sarah is on the trip, a survivor of the dreaded hep.

We plan to carry on in the truck to do the game park circuit, as the company can't get anyone out to fix it yet. Then go to Mombasa for a few days. I will be leaving then for my trip up the Nile via Ethiopia and Khartoum. Poor Geoff finally arrived from Juba isolation hospital.

12 March: Nairobi
Oh yes, we had our final puncture just 10 miles out of Nairobi at the end of the safari around Serengeti/Masai Mara. Dave hitched into town to buy a tube.

20 March: Addis Ababa, Ethiopia
Spent yesterday in surprising misery, missing the crew and the truck. Through thick and thin we had come, and suddenly it was all over. Slept overnight on the airport viewing roof at Nairobi, then flew here; had a great view of Lake Rudolf and the route we had driven down from Juba. Rainy season isn't here yet, lots of trees the same as Kathmandu Valley. Two students took me around the old mosque, although there are mostly Coptic churches here. Haven't seen any other travellers here yet; the locals are very curious and ask lots of questions. Found a dingy hotel with the help of a local fellow. We had dinner together; it cost 32 pence for the two of us. I had chips and Injera, a moist sour pancake. My friend had the sauce stuff – rather hot, it's called Wat.

22 March: Bahir Dar
Met a few travellers here and heard some stories. A New Zealand girl has a graze on her arm where their Land Rover was shot at. Some WHO people had also been shot at. They say it's the

Shifta – local bandits. Still the Blue Nile Falls was a fantastic place, with multi-levels of water and blue-coloured mountains in the distance.

Blue Nile Falls, Ethiopia 1976

26 March: Lalibela
I met an American fellow called Ron and we have been travelling together, which is quite reassuring as we were subjected to a barrage of abuse and stones in Gondar. There was a lot of shooting outside the hotel last night. Anyway we flew for £9 on a DC3 to this place. There are some incredible churches here, cut out of the ground and rocks. One is the shape of a cross from the top and you can climb down to the church. It's called Beit Giorgis; a priest stood outside dressed in robes reminiscent of eastern orthodox but really colourful with the African mix.

2 April 1976: Khartoum, Sudan
Got lucky yesterday. Ethiopian Airlines took pity on me when I tried to sleep at the airport in Addis. I'm running low on cash now, as the money you sent didn't arrive in Nairobi before I left. The airlines put me up in their hotel and fed me, which was just amazing, as I don't like Injera and Wat. It is pretty diabolical after a while. Anyway I'm now stinking hot in Khartoum. The crummy hotel is like a hot house on fire. Just been over to Omdurman across the Nile, you know the Mahdi and Gordon and all that, old boy.

Eventually got a travel permit, so I'm off on the train to Wadi Halfa and then the ferry to Aswan. This is probably my last posting, as I'll soon be up in Cairo. I can get a youth fare from Cairo to London for £53, so I should just make it. See you soon!

****

Well, I got back to Heathrow late at night and slept in car park 3. Hitched to Chichester the next morning and arrived back with 93p left in my pocket.

Trans-Africa again: 1976

After the debacle above, I got a 'proper' job with Trans-African Expeditions. Back to Africa, taking the August southbound overland to Johannesburg. Yes, all that again, only this time we didn't break down much and no one got hepatitis. The trip went very well, the clients were a good bunch. I fell for an Australian nurse.

Beni Abbes, Algeria 1976

The route through Zaire 1976

We routed through Zaire, as Sudan was closed with insurrections in the Christian South against Muslim-ruled Khartoum. This very tragic war continues on to this day in various guises with the new country of South Sudan.

The trip south from Tanzania was all new country for me, with Victoria Falls a memorable stop. I ended up in Jo'burg nearly penniless again, and out of a job, as the next trip north was cancelled.

Before finishing some of us did a quick trip down to Cape Town and into South-West Africa, now Namibia.

Ostrich races, Oudtshoorn, South Africa

I hitched from Jo'burg to Victoria Falls. From Bulawayo to Victoria Falls I was given a lift by a Rhodesian farmer, who insisted I held his gun in full view out of the window. From there to the border, I got a ride with the Rhodesian army in one of their anti-mine vehicles. Later I took the train from Zambia to Dar Es Salaam, before flying back on a cheap ticket to the UK from Kenya.

The flight went via Tel Aviv, and in my youthful ignorance I had bought some bacon in a Nairobi supermarket to cook in the Tel Aviv youth hostel. The warden was not amused!

Trans-African Expeditions closed down not long after that.

The Exodus Years

While I was in the Trans-African agent's office in Johannesburg, I met Dave Gillespie, then a driver for Exodus. He looked rather thin and ill, probably with malaria. They had just started running trips across Africa, having run their first trip in Afghanistan in 1974 and Asia Overland since then. In London one of the directors, David Burlinson, gave me a job.

My first job for Exodus was to build a truck. We worked on the vehicles under a railway arch in Bermondsey, near Tower Bridge in London. There were no toilets except in the café nearby. We usually lived in the back of the trucks under the arch using our own sleeping bags. The clients' seating area in the rear of the trucks had robust plastic used for the roof. The sides could be rolled up on an overland journey. We weren't exactly paid enough to rent a room and in any case we never knew when we would be leaving for another trip. Invariably it would be with only a couple of days' notice, due to some unexpected problem, drivers quitting or being fired, borders shut, visa problems or political troubles.

Every morning we awoke to the rattling of the early trains for London Bridge. The arches were damp, dark, cold and uninviting. Whenever it rained outside, it would continue to pour inside the arch for a long time after the rain had stopped. We ate some of our meals at a nearby 'Eastenders'-style café or cooked under the arch. Is this sort of thing allowed nowadays! One of the Exodus directors, James Wilkins, provided us with a loud radio for music as we toiled away in this dungeon of dark and dank-smelling air. We learnt a lot about engines and how to fix the truck in the dark!

By May (1977) I was on the road, glad to be away from the arches and en route for Kathmandu. We had just four weeks to get the truck out to Nepal with a few paying riders. One passenger was my sister, who was going to Greece to meet her Canadian boyfriend, Phil. It was fun to drive the truck the 7000 miles to Nepal in a month. It gave a better sense of the vast distance. It also gave an impression of retreating in time as each country became progressively more backward. It was, of course, a chance to relive my first trip once more. In Herat I was a little nervous, in case the customs remembered our indiscretions over the Land Rover! But only the young kid at the hotel where we had sold some of the Land Rover equipment recognised me. I spent a lot of time off the streets of Herat in the hotel, doing maintenance or some other pastime, hoping nothing would happen.

Back in Kathmandu, we stayed in Thamel, the new area for travellers. The hippies were gone; a few still clung to their old habits down in Pokhara at the lakeside but most had been unceremoniously pushed out for the coronation of King Birendra. The hippies had now gone to Goa in southern India. In fact I slept in the back of the truck during my stay in Kathmandu, parked outside the now-developing Kathmandu Guest House. We had only just over a week before the next overland departure.

I was to be co-driver to a slick-talking driver/leader by the name of Chris Lee. Chris, the governor, could smooth his way around any problem so long as it didn't involve getting his hands dirty under a truck. That is why I guess I became the co-driver, being by now an experienced grease monkey. I was quite happy to ensure that the vehicle would actually move, rather than compete for a place as top-dog leader. Naturally the most attractive girl among the clients fell immediately under Chris's spell and charming small talk. It was quite a surprise for me too to be the object of some competition among the less pushy females on this Asia trip.

As this trip was leaving Nepal in the hot monsoon season, we decided to spend some time up in Ladakh, the newly-opened part of Jammu & Kashmir state. The region offered a taste of Tibet, with its barren high plateau and monasteries (gompas). The festivals were in full swing; we visited Hemis Gompa and the stunningly located Thikse Gompa, built up the sides of an imposing hill like a mini Potala. Lhasa was a place I really wanted to see, but that seemed a very distant dream with the red Chinese in full control of Tibet and not about to set out on a path of liberalisation. Buddhist Ladakh was a complete contrast to the green, Muslim Vale of Kashmir. The multi-coloured sparkling mountains, the deep blue skies and the Tibetan culture were so different from the rest of India.

Camping on the Zoji La en route to Ladakh 1977

One of the Exodus trucks had suffered serious engine problems and was stuck on the twisting and tortuous road to Leh. The driver was Rob Barton, a leather jacket and denim jeans character, wearing Afghan boots and swaggering about – there could not have been a more stereotypical overland driver, with female clients drooling over his dashing cut. He later went to Poona in South India with another Exodus driver, Renee, and enlisted as a disciple of the Bhagwan. Both drivers dressed in orange robes and became reformed characters; gone was the bravado, the girls and the large ego clouds that enveloped them in their previous incarnations as overland drivers. I never learnt how long this lasted.

Band-I-Amir lakes, Afghanistan 1977

Bamiyan Buddhas, Afghanistan 1977

On our eleven-week overland trip back to the UK we also visited the famous 6th-century Bamiyan Buddhas, which we had missed on our earlier 1974 adventures. Nothing could really prepare one for the impact that these great statues had. It is a tragedy for humanity that they no longer exist. Sadly they were destroyed in 2001, having withstood the ravages of time and nature, but not human intervention.

One day there will be no antiquities, just a world of the same skyscrapers and glass. Further west, high in the Hindu Kush, are the remote and now virtually inaccessible lakes of Band-i-Amir. These incredible deep blue lakes, and the associated crystalline rocks nearby that retain the waters of the lake, are certainly the equal of any great wonder of the world. Will the place ever emerge from the chaos; might adventurous travellers be able to visit the interior of Afghanistan in years to come?

At Goreme in central Turkey, home to the famous troglodyte caves and weird volcanic features of Cappadocia, Chris left the trip to fly back to London for another eastbound trip. Such was the demand for overland trips in the mid-seventies that driver shortages occurred, so I found myself co-driverless and, as it happened, engine-less as well. A piston ring had finally decided to emerge from its grooves and seek a spot in the cylinder head. This had been threatening to happen for some time, as the engine was never overhauled in London. Needless to say, I learnt more about engines and a lot about the skill of Third World engineers. A Turkish welder completely staggered me by welding the aluminium piston and turning it down on a lathe to be almost as good as new, and certainly good enough to complete the remaining 2500 miles to London.

South America: 1977

I still had itchy feet and no intention of being stuck in a job, however good, so in September I was off to South America. I spent two months backpacking from Rio de Janeiro to Argentina and back up the Andes through Bolivia, Peru, Ecuador and into Colombia. Travelling on my own for a longer period was a new experience. I didn't actually meet very many other travellers in Brazil, Argentina or Chile. I don't think South America was really on the backpackers' trail back then. It was a rather lonely trip. I visited the great places of course, like Iguasu Falls and the Inca cities.

15 October: Buenos Aires, Argentina
Spent the last two days in sunny Buenos Aires. There are some fashionable boulevards here like Calle Florida, packed with glitzy clothes shops, very trendy stuff. Been having problems talking to locals, as my Spanish is terrible. Walked down to the Plazo de Mayo to see the Casa Rosades, a pink building where they were changing the guards. Off to Mendoza and Santiago tomorrow.

Buenos Aires, Argentina 1977

23 October: Antofagasta, Chile
Got to Antofagasta in northern Chile at 4.30am in the morning, terrible. Stuck here now until the train leaves for La Paz; it's not a very interesting place. Chile is so expensive, a pound for a meal! I've had enough of this place and buses.

En route to Bolivia
The train to La Paz is incredible, the Orient Express of South America. My compartment is luxury; well, it was once, I suppose. There are two seats of green worn-out leather, there is an old metal radiator but no sign of heat, there is a foldaway sink with brass taps but no water. Outside, the scenery is barren with grey and brown hills. The man sharing my compartment works for the railway. He is a cabin assistant I think, but he's hardly ever in the room. He pops in every so often, takes a large swig of whisky that he has stashed away and off he pops. It's a bit like being in an Agatha Christie novel.

Later at Ascotin
We're at 12000ft, a cool wind. Spent an hour in the 1920s restaurant car talking to some Swiss travellers. We have been passing some smoking volcanoes and deep blue lakes. Saw some red flamingos on one of the salt lakes. A Bolivian girl came into the compartment but we couldn't say much to each other. I think she works on the train. The porter seemed to be smuggling stuff across the border, probably whisky, but he smiles a lot – he's fine really. Some of us have had headaches today with the altitude. The train seems to be stationary for ages in the middle of nowhere.

Next day
We seem to be stuck behind a goods train; it must be the only other train for 600 miles. Passed some great villages with lots of women wearing the famous bowler hats. This is a great place. I saw llamas and smaller ones called alpacas. The air is so clear. We crossed the border last night – no sleep, passports, tickets, luggage checks and all that. Various oddballs keep coming in to bring the whisky man different packets – not my problem, though. Some coaches were taken off and some travellers had to move – strange things happen.

28 October: La Paz, Bolivia
Pulled into La Paz after passing Lake Poopo. The air is chilly but the place has bags of colour and atmosphere. Lots of bowler-hatted ladies, especially around the markets. Found a cheap hotel with some Canadians. There are more tourists around here. Bolivia is great. The old train from Antofagasta in Chile to La Paz was a highlight of the trip. From here to Quito I am back on track with other travellers and revel in some company at last.

30 October: Plaza de Armas, Cuzco, Peru
Really enjoyed the last week, what a place! Got the bus from La Paz to some pre-Inca ruins with fantastic mountains on the horizon. Took a ship across Lake Titicaca, the SS Olantta, built in Hull, England, and carried in bits up to 12,000ft. It's steam-powered, lots of brass and is in good shape. Even the cutlery in the dining car was from Sheffield. From Puno we took a train to Cuzco, more llamas and bowler hats. Today it's festival time in Cuzco. There's a wedding taking place at the main church; everyone is in their best clothing, a band is playing, the dresses are a mass of colours. A flute-player serenades the couple. Candles are being carried forth as a choir sings out. In the distance another band plays from near some of the Inca ruins above the town. What a spectacle! And tomorrow it's Machu Pichhu.

1 November: Cuzco
Another fantastic day. Got up at 4am with some other travellers to get the local train to Machu Pichhu. Very crowded this train; the smell was unreal. We climbed on a switchback and then followed a stunningly green and beautiful valley. A fat lady sat next to me spinning wool the whole time.

Some of us climbed the Huyna Pichhu, the peak that overlooks the main ruins. The path was very dangerous, slippery and wet, and the steps were so steep. One slip and you'd fall thousands of feet to the bottom. The tour groups didn't arrive until much later, so we had a peaceful time. On the train back some Peruvian students played their guitars. Picked up lots of letters at the poste restante, some from Exodus clients on the last trip; they are having problems settling back into normal life, it seems.

Machu Pichhu, Peru 1977

Onwards, ever onwards…
From Quito I crossed into Colombia at Ipiales, a rather forbidding place dominated by brooding volcanoes. In Colombia I got no further than Cali before quitting and taking a plane to El Salvador via an island called San Andreas.

Ipiales, Colombian bus

Palenque ruins, Mexico 1977

Guatemala was fabulous, the colour of the people being the outstanding feature along with the impressive rugged mountain scenery. I was pretty terrified of the bus ride into southern Mexico. It was so crowded with country people with all their wares for a local market. It lurched dangerously from one side to the other. The tyres were plainly rubbing against the bodywork, creating a constant smell of burning rubber. Needless to say the road had no crash barriers and the drops over the edge were quite horrendous.

In San Cristobal I was befriended by a French girl, but as usual I was on a penny-pinching budget and sped on through Mexico to Palenque and passed Popocatepetl to Mexico City. I didn't have time to hang about by now and anyway I couldn't speak French.

I seemed to be on buses forever, crossing Mexico to the United States and the border at El Paso, where nobody spoke English. By now I was reduced to hitchhiking and had some long waits. Finally I was picked up by an anti-establishment, rebellious sort of girl in a pickup truck. As it was getting dark and she wanted to stop for the night, I ended up sharing the back of the pickup with her. I didn't sleep a wink, wondering if I was expected to be a perfect gentleman or not. We travelled a couple of hundred miles before she decided to stop off in the next town. I continued west and spent another sleepless night in a roadside drainage tunnel near Phoenix. Fortunately it didn't rain here much! I hitched on to Los Angeles.

Finally I took a Greyhound bus across the States for $75 and then flew to London from New York on the Freddie Laker Skytrain flight.

Back to the Exodus fold: 1978

Yet again I ended up in England with little cash left. However I still had just enough money to get to Australia in order to keep my Australian residence status going. I needed to be back in Aussie within three years of my last and first visit back in 74–75.

Pitt Street market in Sydney was a psychedelic multi-coloured shopping complex. Each unit was painted in some dreadful bright and lurid shade, and was set up as a mini-stall within the complex. The contract to paint this little lot was run by an Englishman of dubious character. Being paid for four painters but employing only three made him more money and gave us a bigger cash wage. Well, big in respect to English wages at the time.

One lunchtime, while I was taking a break from painting a purple room, I went out into Pitt Street for a sandwich, and guess who came ambling along the street as right as rain. The 'governor' himself, Chris Lee! I couldn't believe what I was seeing. He was down here with the first Exodus Overland group to travel from London to Sydney. First by truck to Kathmandu, then plane, bus and train through South East Asia and across Australia from Perth to Sydney. He told me that Exodus were short of drivers in Nepal and what was I doing wasting my time painting shops in Sydney? By the end of the week I was at a travel agency buying a ticket to the subcontinent with my old Exodus job back.

Having paid for my ticket, I had about £20 left on arrival in Delhi, with a promise of trip funds from Exodus expected at the Thomas Cook office in the Imperial Hotel. I was rather sad not to be starting out from Nepal. Unfortunately the money was delayed, so I was soon on credit at the campsite restaurant, but everyone there was friendly and anyway Exodus were good customers. I got to know the telex man at the Imperial Hotel quite well. He was still working there in 2000. Needless to say, he does not have a telex machine any more, just a swish new business centre with e-mail.

The campsite was as noisy as ever, so more sleepless nights as usual. That's India. I was due to drive another truck empty back to London and this time it was the middle of winter. Dave Mascall, another leather-booted, Afghan coat-clad driver, arrived from Kathmandu and together we shared the long hours of driving back to the west. Dave now resides in Nairobi, looking after lions.

After the Khyber Pass we froze the whole way. We pitched up once more behind the infamous Lodhi Hotel in Kabul, where the manager Ibrahim was, as always, enticing hotel guests with beautiful Lapis Lazuli gems. Afghanistan was as famous for these gems as coats, carpets and heroine. We drank tea by the kettleful and watched the locals smoking their hookah pipes in one corner alcove. I wonder if the Lodhi Hotel is still standing and what has become of the manager?

The road ran south through Ghazni to Kandahar, where a slight rise in temperature was welcomed. Continuing on the loop, the road passes through Helmand and on to Herat. We had to be tow-started, it was so cold before leaving. We didn't dare turn the engine off overnight in northwest Iran and eastern Turkey, and even in western Europe things were no better. Greece was pleasant but also cool.

Return to Afghanistan?
Following 11 September 2001 there was just a glimmer of hope for a return to Afghanistan amongst some adventure travel companies. However, after a brief period around 2004 when a few plucky souls braved the country to explore, it has never looked likely to become safe for general travel. Should it ever become safe, there are a myriad of places to visit, particularly along the central route through the Minaret of Jam, Chaghachuran and on to the Band-I-Amir lakes. To the north is the blue-tiled mosque and shrine in Mazar-I-Sharif, a superb gem of Islamic architecture. South of Kandahar is the magnificent Arch of Bost. Afghanistan was an exciting, if wild country to visit. Will it ever be accessible again in our lifetimes?

Arch of Bost, Afghanistan

Nine months in Africa

Soon I was off to Africa for 9 months, the time it takes to drive down to Johannesburg and back. Living was rough, camping in varied conditions. The southbound trip proved quite a hassle; my co-driver was more aggressive than most and the group was mainly male. Such one-sided groups tended to be more trouble; the best trips always had a good male/female split. We had border problems in Morocco, spent a week in Spain waiting for a direct ferry to Algeria and somehow got into Nigeria. Five of us did not have visas for Nigeria. The embassy in London had also stamped into our passports 'refused visa'. We managed to cover up the 'refused visa' stamps with Algerian postage stamps and put the passports without visas at the bottom of the pile. In true African style a way forward was eventually found and we drove on to Kano. We camped at the back of the Central Hotel and 'Mr Fix-It' Mohamed appeared as he always did when any trucks came into town.

He was now getting fatter since our last meeting in 1976, the fruits of his outstanding initiative. He could be guaranteed to find any Bedford spare part in the amazing sprawl of Kano, as well as change money of course. The old town still retained its Sultan's Palace, colourful indigo dye pits and superb mosque. The city walls were being torn down to make way for the bursting population. The route headed into Cameroon, always a breath of fresh air with its peaceful savannah. Massive termite mounds dotted the rolling hills, shaded by stunted trees and flanked by tall grasses. The Central African Republic was still holding together, but only just.

In Zaire we climbed the Nyiragongo crater that had erupted not long before and since my last visit. (It erupted again in late 2001 and wrought devastation on the town of Goma to the south on Lake Kivu.) And then on into Rwanda.

Climbing Nyiragongo Volcano, Zaire 1978

In Kenya I met Dick Hedges again, the first time since the African trip of 1975. He was a well-known local character who stayed on after independence and ran a safari company. I couldn't imagine him ever leaving the wild beauty of East Africa for the grey of the home country. On this tour we made a side trip to the picturesque Lake Malawi and visited Livingstonia, a piece of old England in Africa, very reminiscent of an Indian hill station. Almost on time, the trip arrived in Jo'burg and I had a couple of weeks to sort out the truck and meet the next bunch of adventurers.

On the return trip north we had a ball. The group was excellent; everyone got on really well. A week's delay while rebuilding the engine in Nairobi only served to bring the party closer together.

There was an Austrian, two Swiss lads, a couple of New Zealand nurses and some Aussies. We had a number of Brits returning from South Africa to the UK after living and working there, including a couple of ex-deb society girls. We often seemed to attract such clients, and it certainly made for a fun trip.

Rwanda

Rwanda was a beautiful country, but the people lived in poverty in the crowded cities. Some of the roads were surprisingly good, a joy after the shocking arteries of Zaire. The devastating genocide of the mid-nineties was no doubt in part due to population pressures, urban poverty and lack of development in general. Today it seems to be going ahead full steam with some political reservations. One big success story is the regeneration of the Mountain Gorilla population as tourism brings in some funding.

Typical panoramic views in Rwanda

We headed north from Rwanda along the densely populated western rift escarpment to Mount Hoya in Zaire. From above the jungle-clad slopes where grasslands existed, a vast panorama was on offer across the dark and forbidding Ituri rainforest. Today Mount Hoya is the domain of rebels and private armies.

Zaire gave us plenty of action; we spent a whole month stuck in deep mud holes, building log bridges and generally getting completely covered in sticky brown mud every day. We were stuck for three days building one bridge and needed to wade through the swamps to get at some trees to make it. Further along we inadvertently sliced off a termite chimney, with disastrous consequences for the night's camp nearby. Living in these conditions was quite a task, camping on mud, avoiding the voracious mosquitoes in the evenings and cooking on wood fires. At times we wondered what the attraction was.

Truck-swallowing roads in Zaire

Building bridges in the third world

Pygmies in Zaire

At the Station de Epulu, where the rare okapi, a large deer, were kept in an enclosure, there are some pygmies. The pygmies here were keen to take passing overland groups into the jungle, showing us their skills as hunters. We celebrated by eating succulent deer and boar.

> ### Aftermath in the Congo
> I often think about those people now, stuck in a terrible civil war. The country has been wracked by armed groups from different factions, backed by money and military assistance from countries in Africa who can least afford the cost of unnecessary wars. Today, as the world becomes more accessible through the Internet, better roads and more information, so it shrinks away as more civil wars, civil breakdown and armed conflict increase. It is harder to travel now than it ever was in the late seventies.

The town of Kisangani has seen fighting since this trip as well. We plundered the local supermarket of tinned food for the next two weeks to Bangui. The store had hardly any food for sale, just cans of fruit or pasta. The boat had not come up the Congo River from Kinshasa for some time and the roads lay in muddy ruins after a particularly harsh rainy season. As we left town we passed a large house on the roadside where a disco was in full swing. A live band played some great music; the Congo style is still very popular all over Africa and these guys were no exception. With masses of flamboyant natural flair, they entertained the colourful throng of black bodies. Shuffling to the rhythm of the African night, the group was invited to take an active part, dancing the night away. We camped in the early hours in an old quarry barely 5 miles up the road.

Another detour was in store on this trip. We had to make an extra-long drive to Niamey in Niger to get Algerian visas. Having come so far west of Agadez we collectively decided with a big nudge from me to go to Mali. We sat it out at the Mali border for a day before the kind official agreed to let us cross the border to Gao. Here other officials would decide if we could have a visa. By paying double the fee this was 'miraculously' possible.

We set off for Timbuktu, following the northern banks of the Niger River for much of the way. All along here sand dunes reached to the edge of the river, a beautiful sight. The driving was fun, very sandy with some dunes to cross. A lot of sand matting and energetic pushing secured the way.

Road to Timbuktu, Mali

Tuareg traders on the Niger River, Mali

The Tuareg villagers on the way came out to greet us; the people were wonderful, some were reticent, others eager to talk. The kids with their fly-blown eyes and dripping noses were always the first to run out.

67

Timbuktu really was the end of the earth. Of course the city was once a great seat of learning, with a civilisation the equal of any in Europe at the time. The ancient mosques were made of mud brick. In the old town were the houses where some of the earliest explorers had stayed, such as Laing, Barth and Caillie. Ringed around the town were dome shaped, matted houses sheltering extremely poor people. They barely kept the all-pervading sandstorms at bay.

Timbuktu street scene, Mali 1978

Developments in Mali
Mali later developed a growing tourist industry, and rightly so, for the country has much to offer, as we found on return trips in 1990 and 2008. When Libya fell apart after the removal of Gaddafi, the arms that were released during the chaos soon found their way to Northern Mali. Conditions were ripe for rebellion, firstly by the Tuareg minority who have been downtrodden for the last fifty years. The situation was then complicated by the Islamists. Sadly today the tourist industry in Mali is really struggling.

We crossed the Sahara via the Tanezrouft route from Gao to Reggane. The truck nearly sank into a saltpan when we were looking for a place to camp near Adra, and we shivered but survived the December temperatures of Europe. We arrived a month late in London and, as usual, had run out of money. The truck had to be towed off the ferry with the batteries long finished. It was then impounded by the ferry company in Southampton until the ferry bill was paid by Exodus in London. Finally a friendly policeman stopped us in London because a left indicator wasn't working. Exodus were very accommodating over these diversions; fortunately the group rated it all great. I barely made it back in time for my sister Christine's wedding.

She married Phil, a Canadian whom she had met in New Zealand when they were both working at the Hermitage Hotel near Mount Cook. She also did an overland trip, travelling through Indonesia and Thailand to Nepal and then by truck to London. Phil did the trip on public transport but came down with hepatitis that manifested in Greece. The wedding day dawned gloriously sunny, with fresh snow on the grounds of Oving village church. A large contingent of Phil's relatives came over for the occasion; most of them had never been out of Canada before. My sister left for Calgary in Alberta, her travelling days over for a long while as she raised two children.

I ended up a couple of weeks later on holiday skiing with some of the group. Back in London work at the Arches was short-lived but very pleasant. I was staying in various penthouses and posh flats with Caroline, a client from the recent Africa trip. She was homeless at this time, having been working as a nurse in South Africa. She moved in well-heeled circles and was offered a different place to crash each week, dragging me, without resistance, with her. She took me to Wimbledon in the summer and later worked in the Exodus office, such is the incestuous nature of the overland business.

Nepal-bound

In March 1979 I flew back to Nepal. I was very excited to be going back. Unfortunately the airline Bangladesh Biman had other ideas. My new co-driver Tom, an affable Welshman, and I were delayed by aircraft mechanical failures in Dubai for one day. At least we saw some of the place before flying on to Dhaka in their ageing Boeing 707. Of course we had long since missed our connection to Kathmandu and had to wait in Dhaka. Eventually the airline agreed to send us to Calcutta and on to Nepal by that route, but the Burmese airline flight from Calcutta to Kathmandu was cancelled due to a festival in Rangoon. Finally we ended up on Indian Airlines the next day to Patna, then finally to Kathmandu. It had taken four days to arrive.

The Exodus base manager was Stewart Wheelhouse, another great character. He could talk his way into a cocktail party at Buckingham Palace. In fact he had to talk himself out of a less salubrious party at the Pakistan/India border on one trip. Suffice to say it was a great story for years to come and remains to be fully told to this day. The Indian customs official seemed to dislike the quantity of pistachio nuts that were stored on the truck. Only the Indians would know why this should be so.

Back in Kathmandu, I took the chance to visit Boudhanath again; the eyes had been repainted for Losar, the Tibetan festival. The traffic was light, bicycles were still more common than cars and Thamel had blossomed into a Mecca for travellers. KC had extended his restaurant in Thamel and had more room. It was always crowded with overlanders. There were at least six companies doing the trip across Asia by now. At the Kathmandu Guest House, Encounter Overland now occupied a small office in the reception area. We had a house in the Maharajganj area north of town, where all the drivers stayed and the trucks were parked and serviced.

Each night we would drive down into Thamel, eating at Jamaly's or the Utse Tibetan restaurant. Sometimes we ate at the Tashi Takgay, the cosy Tibetan place at the Kathmandu Guest House with some nice Tibetan girls working there. For a splurge we would eat at KC's, where the buffalo steaks were legendary. KC himself was always about, looking like a hippy with his long dark hair, entertaining the guests. Sometimes another long-haired Nepali gentleman would play the ancient piano. They never found anyone to tune it.

The westbound trip had another good bunch of punters, the clients as they were known. Kashmir was once again on the itinerary. We always stayed on the houseboats belonging to Haji Razak and his many sons. The main boat was Onassis, which the drivers always got to stay on, it being the cream of the 'fleet'. Dal Lake meant a welcome break from driving and mechanics. All day long the different sellers came by to peddle their wares. It was very relaxing to punt one of the houseboat's shikharas around the lake, drifting below wispy willow trees amongst old tumbledown wooden houses. Then one could row along sleepy canals and across the lake to the Moghul gardens constructed by Jehangir, one of the more peace-loving emperors.

Just about this time the Russians had invaded Afghanistan and no one knew what the implications for travelling through there would be as we left Kathmandu. In the event we managed to get a visa for Afghanistan in Islamabad and continued up and through the Khyber Pass to Kabul. It was a bit of a carry-on though.

Houseboats on Dal Lake, Kashmir, India

19 April 1979: Khyber Inter-Continental, Peshawar, Pakistan
Had some trouble getting Iranian visas but got them in the end. They say the border might be closed, though. Still couldn't get Iraqi visas. Camped in the hotel garden.

20 April: Kabul, Afghanistan
Crossed the jolly old Khyber Pass. Had a bit of a race with a Top Deck group in a London double-decker bus. Got to the border first, which saved us a lot of time, as Penn Overland were there too. At least we could swap stories about visas etc. Drove up the spectacular Kabul river gorge, as always a great ride through the canyons. Kabul seems very cold and unfriendly this time, very few tourists now, half the hippy places have closed. We heard the Iran border was open beyond Herat.

The Iranian Revolution: 1979

Iran too was suddenly in the throes of turmoil, with the Shah hounded out of the country and Ayatollah Khomeini and his gang installed in power in February. The local people seemed very glad to be rid of the Shah and there was a feeling of euphoria in the towns. But looming like dark spectres on the horizon were the Khomeini Guards, a bunch of mostly self-appointed Islamic followers who eventually terrorised much of the country with their increasingly belligerent activities.

In fact they picked us up west of Tehran. They wore no official uniform and were mostly dressed in denims and army-style jackets. Ironically these thugs, wearing almost American-style dress, were those proclaiming most loudly against the 'Great Satan'. We were all hustled into a nondescript room, one block off the main Tehran–Tabriz highway, while our well-armed youthful-looking interrogators found someone who spoke English. This took some time. As we stood in the room, an air of intimidating gloom filled this vortex of anticipation. Guns were shaken angrily at us, and none of us said a word.

Being the driver, I was hauled in front of the most harassing, senior thug. Eventually another man entered the room. The passports were all minutely examined, usually upside down and definitely back to front. 'Ah! Englishman, where have you going?' 'Are you over there, Canada and Australia? Why do you come to Iran?' So many questions. The translator disappeared with some of his cronies and we were made to wait what seemed like all day, but was probably only two hours.

Eventually, unable to decide whether we were emissaries of the Great Satan, or just stupid tourists, they let us continue to Qazvin and so to Turkey and Europe.

Revolution in Iran 1979

Mullahs at the mosque in Qom

A proper job!

I had a big surprise back in London, when I was offered the job of base manager in Nepal for the 1979–80 winter season. This proved to be a very exciting change. It was so pleasant living in Nepal for a whole winter. But first, of course, I had to get there. Another 12-week overland leaving soon in early August. The truck proved to be a rather poorly maintained sample and caused me a lot of late night grubbing about. The whole transmission seemed to be out of balance. It all started in Salzburg on day three, continued in Istanbul, Iran and most days from Lahore to Kathmandu. The gearbox and clutch were hauled off in Nepal for serious reconstruction.

'You, Mr. Infidel, you come to Iran with whisky!' shouted the guard in a green jacket, as he poked his very large machine gun into my stomach. I winced. 'Sorry, sorry for that, it is not mine, it is Turkish arak.' The guard smashed the bottle down on the ground in a great show of anger and distain. Once again I had run into trouble with the Khomeini guards. This time it was at the desolate Turkish–Iranian border. Three clients were refused entry and we were instructed by the Iranian official to go back to Erzerum in Eastern Turkey to modify their visas. Since there were serious fuel shortages in that part of Turkey at the time, I was permitted to enter Iran alone to get fuel.

Unfortunately the guards were hell-bent on turfing out the contents of the truck, and made a very minute and seriously conscientious search of the vehicle. Of course the clients were still on the Turkish side and someone had left a bottle of Turkish alcohol in the bottom of his locker. What a bunch of morons, I thought. It was very unpleasant and I was booted back over the border in no uncertain terms. It was interesting that the customs and immigration people were most courteous and really seemed to go out of their way to be pleasant. One policeman even remonstrated with these guards. It was quite obvious that the regular officials were under the evil eye of the special Khomeini Guards. When we all returned a couple of days later, a different bunch of these thugs were present, and we continued on our way.

On this trip we drove for the first time up the newly opened Karakoram Highway to the stunningly beautiful Hunza valley. Mount Rakaposhi provided a magnificent backdrop to the spectacularly located Mir's Palace. The roof of the palace left a lot to be desired and was badly crumbling away, being constructed of mud set over wooden poles. The road too provided a superb view of Nanga Parbat, known as the 'German' mountain, because it was first climbed (solo) by Herman Buhl in 1953. We camped nearby across from the meadow that would become the trekker's base in future: Fairy Meadows.

Unfortunately my co-driver seemed to think that he was on a paid holiday and obviously had no intention of doing the trip again. Worse still, he was also involved in smoking dope together with some of the clients. This had apparently started in Istanbul and must have been going on through Iran. This was definitely an activity completely forbidden by Exodus, quite rightly, and anyone caught could expect to be fired on the spot. Of course had this been detected by local officials the truck would have been impounded and the driver, me, arrested. Very likely the clients too would have been locked up.

Some of the concerned non-smoking clients alerted me to this in Pakistan and I issued warnings immediately. But one night, as we descended from the Karakoram to Rawalpindi, I caught about seven of the clients and the co-driver smoking in the camp. They could at least have conducted their activities away from the rest of us if they had wanted to. Needless to say I fired the driver and sent him packing in Rawalpindi. I asked the clients to leave the truck. They could rejoin the trip a week later in Srinagar across the border in India. I think they had an interesting time, apart from carrying all their baggage about.

It was a curious thing but when they rejoined, the atmosphere within the group changed for the worse. Not because they had been expelled, but because a couple of others who hadn't been caught probably felt bad. Even the ones whose concern had brought the matter to a head seemed disturbed by the loss of harmony, the group now being perceptibly divided into two sides. Such is the nature of group dynamics when people are living in very close quarters for three months.

Kathmandu: I'll soon be seeing you
A Cat Stevens song

Some of the clients from the overland then joined me for a three-week Everest trek, and this proved great fun. This was my first trek-leading job and a good introduction to the winter to come. This trek route to Everest was a completely different and, at that time, an innovative alternative to the more common route through Lamosangu and Jiri, which the early expeditions had used. We drove first down to the Terai lowlands along the Indian border area of Nepal, and then headed north from a muddy roadhead at Kotari, not far from Janakpur. The route followed a trading route and climbed over the Siwalik Hills, then the steep Mahabharat range to Okhaldhunga. This pleasant middle-hill village was a hive of activity, with porters carrying their heavy doko loads in all directions. Soon we passed through Salleri and Phaphlu, where Hugh had met Ed Hillary in 1975. Most of the party reached Kala Pattar, where some of us climbed further along the ridge towards the cone-shaped Mount Pumori. From here the South Col became visible.

Sherpa Co-Operative

After the trek I lived in the Exodus house in the northern part of town, at Maharajganj. It was a single-storey, simple place with plenty of room and a large garden for mechanical work. By the time I had returned from Everest, some of the other overland trucks had arrived. Most weekends a trekking group would fly in from London, and I organised these with the local operator, Sherpa Cooperative.

The Sherpa Cooperative office, Kathmandu 1979

Mike Cheney, who had been the base manager for Chris Bonington's Everest South-West Face expedition, ran this very efficient agency. The large, successful expedition had put Doug Scott and Dougal Haston on the summit, but they had to bivouac out overnight below the summit on the descent. The BBC cameraman Mick Burke disappeared on the summit area and was never found.

It was not uncommon to see famous mountaineers in the office, for they nearly all relied on Sherpa Cooperative in those years. Mike Cheney had been a tea planter in eastern Nepal, close to the border with Darjeeling in India. Tea is still produced on the hills around Ilam, and its flavour rivals that of Darjeeling. Mike then came to Kathmandu and worked with Colonel Jimmy Roberts, who set up the first trekking agency in Nepal. Later Mike moved to form his own outfit with his old helper Purna Lama, a colourful character who was cook on the Bonington Everest expedition.

Sherpa Coop had an office with a very smelly toilet in Kamal Pokhari, within a walled garden area near the Krishna Bakery. The bakery still existed in the same old house until 2013, but the garden office has now disappeared under the plush Marco Polo Hotel. At least the toilet was well and truly consigned to history, but it is sad to see the garden gone. We had visitors most days, sniffing about for the juicy roots and fresh grasses. A large Brahmin bull was in residence most of the winter, along with other holy cows of his fancy. I bombed about on my Indian Hero bicycle between the house, the office, Thamel and the Hotel Blue Star down in Baneshwor.

Time flew by; Christmas Eve came. Encounter Overland had their party. I drank too much rum and coke and missed most of Christmas with terrible stomach pains. I didn't take well to the booze. I met Mr Singh, the fortune-teller, at the party and he too seemed to be suffering from the booze next day. There was a very friendly atmosphere between overlanders and trekkers. Not far from Thamel in Jyatha, a Sherpa suburb in Kathmandu where the hill people from Everest/Solu Khumbu stayed in the trekking season, was the Sherpa Expeditions office run by Sherpas Pasang and Chotri. Ben Wallace, an American, worked as the base manager. New Year parties were held there most years.

The Encounter manager was none other than Martin Watkinson, whom I had first run into when he was a driver. We had met in the middle of a muddy hole in central Zaire back in 1976. By New Year all the trucks had arrived from London and disgorged their punters. The drivers were a great bunch. Roger worked diligently on the beast of the gearbox from my trip. Henk was a large, affable Dutchman and, being a trained engineer and mechanic, could fix anything properly while some of us were still learning the finer points. Phil Normington had a Ph.D. He had the dubious reputation of having suffered the periodic Exodus ritual of being fired and then reinstated soon afterwards. He later became the only member of our group to stay at Exodus – until 2014. He became more office-bound and of course is now old and mostly out to grass! Howard was a teacher from a far corner of the west of England and wasn't a mechanic at all, but made up for this with boisterous laughter and entertaining gossip.

Alan was hired for the last trip westbound when the company, Treasure Treks, he had been working for went bankrupt. It was a messy affair, with drivers not being paid and trucks being bought and sold. I also met Stan Armington, an American running a trekking company and a subsequent writer for Lonely Planet. When Treasure Treks went broke they also owed Stan some money. He managed to get one of the trucks impounded and hidden somewhere in the valley. It was a truck that Exodus had bought back in London. I was to be in some conflict with Stan, although it was all quite amicable when I finally tracked him down hiding from me behind some very tall cabbages and cauliflowers in his garden.

Sir Crispin arrives

In the spring the British army arrived in the form of one Sir Crispin Agnew of Lochnaw Bt and his climbing expedition. We were engaged to ferry them and their compo rations, ropes and gear to west Nepal – the objective to climb the isolated and little-visited Mount Api.

The logistics for this expedition proved almost as difficult as the climbing. Phil Normington and I set off with two trucks and all the lads for Dandeldhura, the roadhead close to the western border of Nepal and India. In those days there were no roads to the west. Our route was to be via India, through Gorakhpur, Lucknow and back into Nepal near Dhangari. Unfortunately there had been a fuel crisis in India and thus Nepal for most of the year and we could only get 50 litres at a time. To get fuel entailed a visit to the District Commissioner or some other official to requisition the correct bit of scruffy paper in order to get the fuel. Long before the expedition, and indeed before the spring overland trips, we had been hoarding fuel, going through the ritual of obtaining the said papers every three or four days. Each time a different driver had to do the honours. We also had some very large fuel tanks made to do the job.

Once in India the fuel problems continued. We spent a day in Faizabad, near Ayodhya, getting a permit and fuel. The army lads relaxed in the garden with Sir Crispin dressed in all his finery. His kilt caused a lot of amusement with the passing locals. In Lucknow we again were confined to the gardens of the Gomti Hotel with fuel delays. Phil and I went off to the Mayfair restaurant for the best Indian food in town. Next day we drove northwest along dusty brick roads, with badly broken tarmac strips balanced on the sinking bricks. We crossed a large river and ploughed through sandy tracts to reach the border. Our roadhead the next day lay up some very tortuous roads with some very bad avalanche-prone sections threatening to block the return trip. The army lads unloaded tons of compo rations. Although invited to join the march to Base Camp, we had work back in Kathmandu. We bade them farewell and good luck.

> ### Ayodhya
> Some years later in 1992, Hindu zealots razed the Babri Mosque in Ayodhya to the ground, brick by brick. The mosque was said to occupy the site of a temple dedicated to the Hindu god, Rama. Communal riots broke out all over India after this act. This problem marked the beginning of the rising tide of Hindu nationalism in India. The traditional secular politics were slowly being eaten away. The Bharatiya Janata Party (BJP) replaced the Congress Party in power. This blatantly Hindu party had a different, more religious-based agenda. After that the spectre of communal disturbances was never far from the surface, until more recently. And all this has changed since the Sikh trouble in the eighties and Kashmir in the nineties. Such was the pace of increasing intolerance in all parts of the world, even back then.

The expedition made it to within sight of the summit, but was forced off the mountain by bad weather. Poor Sir Crispin suffered a nasty eye injury when shards of ice crashed on to him. He later recovered.

On our return, just before Lucknow, we ran into a hostile bunch of drunk-looking locals bent on blocking the road. For most of the day the road was ominously empty, this being the middle of the Hindu Holi festival. In Kathmandu this festival had always been a light-hearted, almost childish, affair, with gangs of youths and others throwing coloured water and dye at any passer-by. Here in Uttar Pradesh, a poor area of India, things seemed much more threatening. Finally a stone came through my windscreen and we drove in haste for the calm sanctuary of the hotel coffee shop in the Clarks Avadh hotel, where we took refuge until dusk, when the over-boisterous festivities concluded for the day. There was no opportunity to visit some of the fine mosques and the Residency, where a mutiny had broken out in 1857 against the British, a siege of a different kind. It was a great relief to see the twinkling lights of home the next night as we raced back into Kathmandu.

Soon the trucks began to leave with their new groups, mostly Aussies, New Zealanders, and Brits from Australia. Howard left first, then Phil and Roger. Henk found a new job working for Hann Overland, being more interested in staying longer in India. The trekking season was also coming to an end, with the hotter weather of late April. During the season a number of clients had found me more attentive than the job required, but I had only fleeting time with most of these young ladies. There was a certain element of competition among the lads for the girls and everyone was naturally interested in the first meeting with their new groups. I got to meet many of them at Sherpa Cooperative when they first reported. Who would be sharing the trip with whom? Such gossip was bound to be the talk of the night in one of the Thamel cafés. I wasn't down to do any overland westbound and was to go to Kashmir with the equipment for the coming trekking season in Ladakh. Just before I was to leave, I met a pleasant, somewhat cheeky, young lady who had been travelling around India with some girlfriends. One had been given my name by someone in India. Melanie came from Dorset and had left York University fairly recently. Her friends were leaving from Delhi soon, but she decided to meet me in Delhi and come to Srinagar on the truck.

By now the weather had become stiflingly hot in the run up to the monsoon, with temperatures approaching 40°C in Delhi. We quickly headed for the cooler climes of the Kashmir Valley. This was like being on holiday, although I could hardly have said that the Nepal job was a proper job. The winter had been great and it ended on a high with Melanie. I flew back to London just after her.

Sudan or bust

An Exodus truck had crashed in Kenya and had been written off. A very old Bedford had been hurriedly bought to replace the original truck. That too had suffered a close accident, when the brakes failed on the road out to Masai Mara. Most of the clients had abandoned the trip in anger, but some insisted on continuing the journey. I was flown straight out to Nairobi to sort out this little lot. It was quite a shock to be back in East Africa, camping at the Equator Inn. Exodus agreed to let Melanie accompany me, as there was no co-driver.

Chris Lee was in Nairobi with a group. He was in fine spirits, most of them slipping down his throat. He was very thin, as we all became after some months in Africa. Unfortunately Chris had written off a truck in Iran, with a lot of serious bureaucratic hassle. Fortunately no one had been seriously injured. He took a shine to Melanie, of course, but the booze drowned out his smooth talking. I set about fixing the brakes.

The trip was a disaster, the truck was dreadful, a wreck; the clients were quite aggressive, but that wasn't too surprising. We drove through the game parks to Lake Victoria. Then a passing bunch of wildly disciplined Tanzanian soldiers scared us out of our wits. They had recently been sent to Uganda to keep the peace. The presence of these military yobs prevented all sleep for a couple of nights, as we tried to hide away in the bush.

Terrible roads and bridges in NE Zaire

The roads into Sudan from Zaire were atrocious; some bridges had collapsed and others had no wooden planks left on them. Somehow we had to drive across these obstacles, using sand mats and the metal rails that remained. A lull in the fighting between the government in the north and the rebels in the south of Sudan enabled us to proceed to Juba. Massive granite boulders littered the landscape of savannah bush near the small settlement of Yei. Juba had not changed much; there were still no proper eating-houses.

We had hoped to continue up the Nile to Khartoum and Cairo by way of Wau and El Obeid. In the event this was not an option, as the truck engine blew a head gasket in the main square of Juba. It was very fortunate that it happened in the only town of any size, but the truck was plainly knackered and was donated to a passing aid worker. I kept the carnet, he kept the keys, and a deal was concluded by Exodus sometime after I returned to England – I think. Suddenly I was out of favour at Exodus.

I did another trip to the Middle East with a combined Exodus/Encounter group, who had to fly over Iran from Amman to Karachi. Things had deteriorated badly in Iran by now, and no one was able to drive through for some years. I was back in England soon enough, after a week of solid driving back empty from Amman in Jordan. Melanie and I were getting on pretty well. The winter passed slowly as I worked in the new Exodus workshops down in Pewsey, building a new super-truck for the first South American Overland. Pewsey was a vast improvement on the old Arches.

South America Overland

The angry stranger, his eyes bulging with anger, tried his best to open the rear door. Our sleep had been rudely disturbed by some shouts. We awoke in horror. The man seemed to be about to break the window. What the hell was he intending to do? Melanie buried herself under her blanket as I hurriedly scrambled out of a side window. Melanie also scrambled out and we ran for a house down the road. It was locked and bolted. We quickly ran for cover in a ditch. The man seemed distracted by our thrashing in the black of the night. As he stumbled down the road looking for us, we ran back to the truck, I fumbled for the ignition keys and tried to start the vehicle. Blast the thing, it didn't want to start in the cool high mountain air. We were at over ten thousand feet. After some heart-stopping moments, it fired, and away we drove, on and on throughout the rest of the night until dawn. At first light we stopped in a roadside café, drinking copious cups of strong coffee in an attempt to recover some semblance of normality. We were about thirty miles north of Medellin in the central highlands of Colombia.

The truck had been shipped or slipped out on a banana boat from Liverpool to Cartagena on the Caribbean coast of Colombia. I had flown out to Bogota, arriving just before midnight. The guidebook said specifically, 'Don't get a flight into Bogota at night.' Scared half-witless, I emerged from the airport and took the first taxi I thought was okay, but then how could you really know? The guy hurtled off and took me to a different hotel from the one I had requested. I didn't argue, and anyway I was still learning Spanish. It was such a relief to get into the hotel. Of course it was more expensive than the one I had asked for.

The next day Melanie arrived and we flew to Cartagena. Then started five days of hassle with the customs and our Mr Fix-It agent to clear the truck from the port. Cartagena is a gem of a place, a traditional Spanish town with a lot of very pleasing buildings, churches and squares. The customs were less pleasing and demanded a lot of money – 'gifts' they said.

On the road to Medellin we were stopped by officials at the many check posts. All, without exception requested money before we could proceed. Fortunately the amounts were mostly relatively low, a few dollars more here and a fistful there. We drove for hours in the mountains. The lower slopes were green and lush with coffee or banana plantations. It was hopeless trying to find a suitable place to camp, so in the end we just pulled into a lay-by. It was here that the angry man gave us such a fright. In the calm of the morning we suddenly realised that the man had spoken some English as he ranted and raved in the dead of the night. We will never know what his problem was. Drugs must have played a part.

Bogota was a place that grew on you. The streets were blue with the haze of petrol smoke; the noise of the vehicles was deafening, no one believed in car silencers. Modern skyscrapers jostled for position in the central business area while out in the poor suburbs people lived in cardboard boxes. Melanie had paid a contribution to Exodus to join the trip, but was hoping to get some photographs, material for her portfolio. She had a keen interest in becoming a photojournalist. One afternoon I took the truck into the mechanicos area to do some welding on the roof seat. We were in one of the poorest areas of Bogota. We had nothing but help and kindness from the people working in these garages. Melanie visited a nearby slum to take some pictures. Again the people were courteous and accommodating. In one area some children and their parents were watching a football match on a television. The set occupied pride of place in this collection of cardboard box homes and tin shacks.

Later in the week, Steve the new co-driver arrived. We went to the bank to collect the trip funds. Stalked into the place by some street hustlers, we sat in the bank for over an hour hoping they would lose interest; luckily they did. We scurried back to a café to eat the plate of the day and listen to some great cowboy music playing on the radio.

The trip was twenty weeks long, the first trip for Exodus in South America. I had the exciting job of leading this first foray into a new continent. My time in South America in 1977 had brought unexpected opportunities. It was a fascinating experience; the trip visited a salt cathedral on the way to Venezuela. We crossed the Gran Sabana, a vast, forested upland set like an island surrounded by even denser jungle. Mount Roraima appeared in the gloom and for a couple of days we were prevented from continuing by a raging river torrent. The incredible size of the waves of this cascading water was astounding. It rained a lot.

Climbing the Gran Sabana, Venezuela 1980

Mato Grosso country town, Brazil

In northern Brazil the road through Boa Vista to Manaus was awful. The famous opera house here was a fading relic of the boom days, when Manaus was opened up for exploitation of the rich resources in the area. From Manaus to Porto Velho the road passed through reservations belonging to untamed Indian tribes. Some bright spark among the punters regaled us with stories about the natives here, who hunted with poison-tipped arrows. No one wanted to sit on the roof seat. The road was not sealed all the way and we had to cross a number of large rivers on decrepit ferries. If anything, the road was worse through Mato Grosso.

Brasilia was an intriguing place, a better example of a new purpose-built capital. The traditional Portuguese-style picturesque town of Ouro Preto was perhaps the nicest town in Brazil. In Rio I spent a lot of time at the airport customs extracting the spare injector pump that finally arrived from London. No lounging about in the cafés of Copacabana on this trip.

On the road in Paraguay 1980

Steve was an amicable co-driver and very capable. He later worked for Hann Overland in India, then went to live in Thailand and have a family. We headed south through to Iguasu Falls. In Paraguay there was dissent in the group. Some wanted to stay in better hotels; some were on tight budgets. This was one of the problems with such a trip, where clients paid for their own hotels. Steve or I decided on the camping places.

South of Buenos Aires we hit the freezing winter of Patagonia. Along the way were small Welsh settlements of hardy immigrants, eking out a living from the poor soil and offering cups of tea to the occasional passing tourist. One morning we awoke to snow and a frozen engine. Even the brakes had frozen. Fortunately a passing truck stopped to towed us to the nearest café and help us warm up the engine, using some wood from the café owner's stock.

The whole gang went skiing in the ritzy resort of Bariloche in the shadow of Mount Tronador. As we headed north from Puerto Montt in Chile through Santiago to Antofagasta, it warmed up again. Chile is a very long, narrow country when driving its entire length, across varied terrain from rich cattle country, through rolling pine-scented hills to arid desert.

One side trip before the Bolivian border was startling – a trip to the vast opencast copper mine at Chichiqamata.

The copper mine at Chichiqamata, Chile

Crossing the Andes from Chile into Bolivia

The road across the Andes between the smoking volcanoes, lakes and saltpans was as exciting as the rail route I had taken in 1977. San Pedro de Atacama and the brilliant white saltpans at Uyuni were the high points. It was amazing to drive across these saltpans. One could only follow the faint tracks by wearing Polaroid glasses that picked up the variations from pure white flats and the disturbed markings created by vehicle tracks.

In La Paz Steve was attacked by a dog and had to go through a series of painful rabies injections for the next few days. The mountain roads in Peru were truly tortuous. One day it took all day to cross a monumental canyon. Our camp that night was virtually in sight of the previous night's. Everyone loved Cuzco and the Inca cities. Cuzco was a bit like Kathmandu; all sorts of characters seemed to be hanging around. It was a crossroads for travellers. The food was good too. Ecuador enthralled everyone with its colourful markets and brooding volcanoes.

Cotachi market seller, Ecuador

Riobamba market ladies, Ecuador

Soon we arrived back in Bogota, where Steve surprisingly quit to work for Geoff Hann. Melanie and I flew back to London. Alan Alcock flew out to Colombia to do the second trip, armed with a book-load of notes from me.

San Augustin, Southern Colombia

Another winter in Nepal

After one week in England, I flew to Kathmandu for another winter in Nepal. I was to take over from another Exodus character of some fame. I first met him in a pub in London Victoria back in 1978. I always remember him asking me why I was doing the job, why wasn't I doing it for the money? He was a master at making money on the side; he regarded the wages as pocket money and dabbled in all sorts of deals along the way. His most famous exploits usually involved selling whisky in Pakistan in large quantities to officials, hoteliers and anyone willing to pay the price. His notoriety increased when it was rumoured that he had substituted cold tea for the whisky and used resealed Johnny Walker bottles from a hotel.

Of course my answer was that I didn't do it for the money. I suppose I did it for the adventure, the challenge and maybe the temporary sense of importance. It never translated to confidence in normal life. I don't really know why this job that was going nowhere was so addictive. Perhaps I was subconsciously seeking the 'green-eyed yellow idol', the image of something beyond, something more to discover. I wasn't into selling more than two bottles of whisky to pay for a few odd expenses. I didn't spend much and gained an equally famed reputation for being tight-fisted, even with the Exodus money.

My second season in Nepal was equally interesting, although there were more problems. I never seemed to have enough funds from Exodus. Times were harder in the overland business after Iran and Afghanistan changed so radically. Telex was still the only way of communicating, as phone connections were pretty bad. This was a blessing, as it meant there was little interference from the bosses in London, but there was not much cash either. I got to know the man at the telex office in the Shanker Hotel in Kathmandu, as well as the man in Delhi at the Imperial Hotel.

We didn't have the house any more, and I was to be staying in a hotel in Thamel. This was quite nice actually, because I could easily eat out in Thamel without the hassles of cooking. I ate a lot in the Kathmandu Guest House at the Tashi Takgay, which had now changed its name to the Astha Mangal. Here Tashi, the manager's younger sister, was always the star attraction for the drivers.

She had met Chris Lachman while I was in South America and they were getting married. Chris worked for an Australian trekking outfit AHE (Australian Himalayan Expeditions), with a lot of business. I never tired of the congenial surroundings of the Tibetan restaurant, which was being managed by Ugyen and Tukten, both Tibetan exiles from Darjeeling. They eventually managed to get factory jobs in Australia through one of Chris's mates. Ugyen went on to be a driver at the Nepalese Embassy in Sydney.

Melanie was busy in London working on her new photography projects and trying hard to find a way into the difficult world of photojournalism. She stayed in England until the New Year. I didn't get out into the hills of Nepal much until March, when I managed a quick trek up to Jomsom.

The troubles in Iran and Afghanistan had substantially reduced the demand for overland trips; I had only two trips to prepare for and only two drivers. One, Alan, was an Ulsterman and the other was from Rhodesia. Alan contracted hepatitis and had to be flown back from Delhi after only a week or so. The Rhodesian took over but he came to grief in southern Iran, crashing badly. No one was killed, but some sustained nasty injuries and the trip was abandoned. It was a bad time. Jimmy Carter had sent in the troops to rescue the hostages from the American Embassy in Tehran. It failed badly and Iran again closed its borders. Encounter Overland, Top Deck, Contiki, Penn Overland and Sundowners all had vehicles stuck on the wrong side of the border, causing severe complications, with vehicles having to be shipped back to the UK from Bombay.

On a lighter note, Melanie had arrived, and we took a week off to visit Darjeeling and explore some of Sikkim. She was getting restless, though, and after two and a half months returned to England to continue her career ambitions. I was coerced into another half an overland trip, taking the last truck from Kathmandu with the genial and occasionally inert Chris Troman as co-driver. In Karachi the group flew over to Jordan. I was asked to stay on in Pakistan, but by now, after almost two years without a break, I was pretty jaded and tired. I had decided somewhat half-heartedly to quit. I couldn't keep on doing overland trips and still hoped to stay together with Melanie, so I flew back from Karachi. Steve Dallyn was to fly out to replace me. Amazingly we had both been at Chichester College in the same class but I hadn't seen him since.

When I was refused a visa for Iran and couldn't go to sort out the mess of the crash there, I decided to quit. It was in fact a lucky escape from the turmoil of Iran. Things got worse. I hankered after a job in Nepal, but Melanie was getting restless with the travelling life. She ended the relationship soon after I quit my job! Out of work, out of salts and out of luck, it seemed.

But not for long…

I was unexpectedly scooped up by Keith Miller at Sherpa Expeditions for a few weeks in the Alps putting up tents, buying food and driving the baggage and camping gear around on the summer Tour du Mont Blanc of 1982. This proved a completely different style of 'adventure travel' but was surprisingly fun. I met Keith Miller's wife Shanan, leading one of the groups, as well as Kay White, another stalwart Sherpa leader. The walkers proved very enthusiastic and this was to be the beginning of a long association with Sherpa through ups and downs.

Later Geoff Hann, who ran overland bus trips around India, came to my rescue and I found myself in the Delhi YMCA to take over from a driver who had done some dodgy deals with Geoff's trip funds. I met Eugene Pram in Jaipur, who helped fix the ailing bus and continued to Kathmandu with a great bunch of clients. I did an exciting trip from Bombay to Delhi as well as another to Nepal.

After Christmas I flirted with starting my own trekking company. I spent a few weeks in Kensington in London as I started the business. To fund some of this I found employment as the front desk manager in a hotel run by an Iraqi friend of Hugh's (from the 1974 overland trip). Apparently a Palestinian businessman was also a partner. It seemed to cater to rich oilmen from Nigeria and the Gulf. The Malaysian female receptionist had a few rum stories to relate about some of the activities of the guests.

I hated London and my itchy feet soon put to one side all thoughts of the trekking company. It didn't help when Geoff Hann offered me another job. Yet again I found myself in Delhi, this time meeting John Meyer from Exodus. We drove a Hann Overland bus to the border of Pakistan and John returned to Nepal. John had his pride and joy in the bus; the infamous Royal Enfield. It kept burning out piston rings later, but on this occasion lasted long enough to get us to dinner in Amritsar. The bus was in a bad way, so I had to fly to Quetta and take a train to the border at Taftan to meet the group. We took a private minibus to Quetta and then various trains to Lahore. The bus had been fixed in Trevor Lee's garage in Lahore and the rest of the trip ran relatively smoothly.

After this, back in England I was in demand. This time it was for Sherpa Expeditions again, leading a couple of new treks in Kashmir and Ladakh.

Crossing the Lonvilad Gali glacier, Kashmir to Ladakh trek

An overlander, maybe!

PART TWO: A COUPLE FOR THE ROAD

Siân... from Moel Siabod through Oxford to Nepal

I suppose it all began when my father took me to climb my first mountain at the tender age of three. The peak was Moel Siabod, a hill in North Wales, but for me it was the start of an adventure and a passion for high and wild places. Throughout my childhood, living from the age of six with my brother and sister in south Derbyshire, we visited our grandparents for holidays in both North and South Wales, so I was never far from the mountains.

At the age of fourteen I was asked what I wanted to be when I grew up – it was clear to me that I wanted to be a dancing explorer. I was quite keen on ballet at the time, but that soon faded. However, the urge to explore has never gone away and, I dare say, never will.

I remember drawing a map of Australia, a continent I had until 2015 only visited for one night en route from New Zealand to England. There are always more places to go...

Although my university days were spent in Oxford, with college roofs and spires the only climbing locations for many miles, at least once a term we would head off for the hills or crags of the north. I learnt to climb at the OUMC, with Stephen Venables, Phil Bartlett, Roger Everett, John Weatherseed, Malcolm Allitt, Chris Lewis and Dave Oliver, among others. There were only three women in the OUMC.

In Easter 1975 we hitchhiked up to Fort William to climb on The Ben (Ben Nevis). Women were not allowed to enter the hallowed confines of the CIC Hut, so I was obliged to camp outside in the snow, snugly ensconced inside two sleeping bags, with Stephen Venables on one side and David Lund on the other.

Siân punting at Oxford & climbing on Ben Nevis

On later trips, as one of the few who already had a driving licence, I was usually one of the two drivers of the university minibus. Little did I dream then that 25 years later I would be driving my own bus all the way from England to the fabled city of Kathmandu!

Siân climbing on Thor's Rock, The Wirral

In my mid-twenties, I remember one day in Teesside, standing in a fitting room trying on some new office clothes, looking at myself in the mirror and feeling it was not me at all. I couldn't exactly define why, but with my hairstyle, very short and severe, I just didn't feel it was me that I was looking at. I felt as though I was in the wrong body.

I had always expected to continue in business, to lead a conventional life, but it just didn't feel right. After university, I moved into computing, working first for Unilever and then British Steel. From here I did manage to travel, but it was still within the field of computers. It seems now like another incarnation, working in The Hague for Shell Nederland Informatieverwerking, and then New Zealand Farmers Fertiliser in Auckland. New Zealand seemed worlds away from the rest of the world – indeed it still is, and that, for many people, is its major attraction. But for me it was still not right and, on my return to Britain, I had my first experience of Nepal.

That was it. My heart was trapped, and I had to go back as soon as possible. Returning to England to start a new life, I put aside the money I would need to go back to the Himalaya before finding a new job – in computers of course, for where else could I look? Starting at Datastream, in the City of London, I wore a smart suit and travelled on the North London Line along with many hundreds of commuters. Life in the city is certainly a rat race, materialism taken to extremes. The job did not work out as expected and I soon moved to NCR, where I joined the sales support team. I bought myself a small terraced house just outside Richmond, furniture and all the trappings of city life, including a car. But still something was missing. Every day I would wait for the train at Kew Gardens station, looking upwards at the jumbo jets as they came into land at Heathrow, wondering where they had come from and where they were going... Why couldn't I be up there too; was I searching for the impossible?

The time came to choose my summer holiday, and I made a decision that would change my life forever. I had contemplated going to Corsica, because it was much cheaper and still had mountains. But the Himalaya had truly captivated me, so Corsica had no chance. Little did I know that, by forsaking Corsica and choosing Kashmir, I would meet the love of my life, a fellow soulmate who would share the same ideals as I did. I still haven't been to Corsica, and neither has he! It is now somewhere we will go 'when we're older'!

Kashmir is a wonderful place, but as I write this it is being torn apart. Politicians pontificate, thousands of soldiers are lined up on either side of the border and the world's media is crying 'Nuclear War!' In 1983 it was a blissful place, with attractive lakes, strikingly beautiful mountains and hospitable people. We floated along the byways of the lakes in our private shikhara, with willows hanging gracefully down, just glancing off the surface of the waters. Exquisite pink and white water lilies bloomed everywhere, just as they have done since the time of Buddha – there couldn't be a more romantic place to fall in love.

But he was the leader, and I was just one of the group members, so much discretion was required! Our flight from Ladakh to Srinagar was cancelled, because Rajiv Gandhi had commandeered the plane for electoral purposes, so we were forced to return to Srinagar via Kargil, the town on the Line of Control, which is periodically in the news, and some spectacular mountain passes. The winding roads did not agree with my stomach, and I remember spending most of the time next to Bob in the back of the cramped vehicle.

After this trip I returned to England, where I tried to fit back into London society. It didn't work, of course, and every so often a blue airmail letter from the East would drop on to my doormat, lifting my spirits and enlightening my dreams. At Christmas Bob returned, briefly, for the day after his arrival he was offered another job in India. I soon realised that if we were to have any hope of maintaining our relationship, I would have go out there too! Two weeks later he was gone, on one of those buses. In those days there was no email, and telephone lines were almost impossible to get in India, so our only contact was those little blue scraps of paper.

Bob continues…

After Kashmir I had decided to visit China and the Far East and Japan before returning to Nepal to lead a couple of treks in the Annapurnas for Sherpa Expeditions. I could hardly afford to eat anything in Japan and the hostels were pretty expensive. The highlight was Kyoto and the friendly, helpful people.

Kyomizo temple, Kyoto, Japan 1983

On the Taiwan stopover there was just time to see the museum of Chiang Kai Shek.

Taipei central monument, Taiwan 1983

Independent travel in China had only been possible for a year or so and I was keen to get going while the chance existed. There was an odd system for getting visas in Hong Kong, but no snags. I crossed into China on the train to Canton, visited the famous rock scenery of Guilin and took a train to Chungking (Chongqing). Train travel was diabolical if you didn't get a hard sleeper. Sometimes you did, sometimes you didn't.

Teatime in Chongqing, China 1983

Chengdu, China 1983

It was amazing to be in such a wild and remote part of the world, having never dreamed it would ever be possible. I got as far north and west as Turfan and Urumchi in Xinjiang province. At the time Turfan consisted of one street, a great mosque and lots of colourful sellers offering the famous Hami melons. The long march east took several days to Xian, Beijing and Shanghai. It was a hard trip and a lot of time was wasted because of language difficulties.

China before the great leap forward to modernity

I had a lot of time too to ponder on what might develop with Siân. She met me just before Christmas at Heathrow after the Nepal treks, and I stayed a few days in her terraced house in Richmond.

Addicted to India: 1984

After Christmas in England I got the call from Geoff (Hann) once again – a trip around Rajasthan. It sounded too good to miss, all those palaces, forts, colourful people and bad roads. Jaisalmer, Jodhpur and Udaipur were the obvious highlights, but the lesser-known places such as Mount Abu and Ranakpur, with its delicately carved marble Jain temples, were equally impressive. You could be sure to see some great backroads on a Hann Overland trip, as well as lots of dusty mechanics' yards and little sleep in the back of the bus. The trip also included a few days in Goa, as well as old favourites like Ellora, Aurangabad and Dhulia. En route was Poona, famous for the original antics of the Bhagwan. It must have an amazing influence on visiting foreigners.

One could always be guaranteed some interesting clients with Hann Overland. This was no exception; we had a couple of gay gentlemen, a teacher from north London whom we met years later on a bus from Derby to Victoria, and various others including a young Dutchman and a rather large lonely Scottish lady who was nonetheless very brave to be on such a trip. Unfortunately she seemed to think that the driver was fair game, and made her intentions quite plain. Initially things proved fine, but in Poona there was somehow nowhere to hide. Of course the bus maintenance took up a lot of my time, but in Poona it took up so much of my time that I was getting a stiff back lying under the bus for hours. This I decided was the only safe spot to hide from further amorous advances. After me, the young Dutchman seemed to be the next candidate. In Poona the past was orange, the present was red hot and the future…!

Ranakpur Jain temple, India

The clutch finally gave up just outside Bhopal in central India on the way north to Delhi. I had hoped to visit the Buddhist complex at Sanchi, but had to spend the whole day on the side of the road fixing the problems. At least I managed to buy a genuine Bedford clutch this time. It must have been so; it was printed on fine-looking paper. Well, those Indians, they know every trick in the book when it comes to business and survival. I always thought back to the man who sold me that genuine imitation clutch in Bhopal. I wonder if he died along with thousands of others, when the Union Carbide factory blew apart some months later.

The first stupas were believed to have been built in Sanchi by the Emperor Ashoka in 3rd century BC. I never did see Sanchi, but the clients said it was a thoroughly interesting excursion. Fat lot of good that was for me, but at least they were happy.

There was a long letter from Siân in Delhi. All I had to do was to drive the bus to Nepal, park up and fly home to see her.

March 1984: Siân

In March Bob returned to England for another fleeting visit, during which we planned our next venture. I would fly out to Karachi and on to Quetta in the Baluchistan desert on my own. I would meet Bob there and we would continue up to Skardu and the Karakoram. All the plans were made, but in the end we decided that if I was going to give up my job in August, it would be ridiculously expensive to fly out to Asia twice.

June 1984: Bob

Because of the political situation in the Punjab, while I was in the air on my way to Delhi the Indian government instantaneously introduced a new law requiring all foreign visitors to have visas, even those already in the air on their way. Fortunately I was due to fly on a connecting flight to Kathmandu to collect the bus, but other passengers flying to Delhi were forced to fly to Kathmandu or Bangkok to obtain a visa!

Baluchi men in a teahouse, Pakistan

Local truck, Jacobabad, Pakistan

Once again the Hann job involved a horrendous trip on public transport in Pakistan. This time things were rather more serious. The Punjab was in flames, with the Sikh rebellion in full swing. Sikh rebels had holed up in the Golden Temple; their leader was Sant Jarnail Singh Bhindranwale, a kind of folklore terrorist. They were demanding a separate state from India. I had already repaired my bus at Shafiq's in Varanasi on the way to Delhi, but the Pakistan border was suddenly firmly closed. I had to return to Kathmandu and park it there, because Indian customs did not permit vehicles to be left in the country without a driver attached. This was only possible in Nepal. I then had to fly to Karachi as quickly as possible. The group were expected in three days in Taftan, at the Iran-Pakistan border. It was last year all over again. Disruption on the Asia Overland trail was at its worst since the Iranian revolution.

At least I got to see a different part of Pakistan this time, taking a bus from Karachi through tribal Kalat to Quetta. The journey in a truck to the border was no better than the train I'd taken the previous year. I was absolutely on the ragged edge when I arrived at the border and slumped in the shade of an empty mud shelter. I fell asleep. Little did I know that the clients had already crossed the border with a new trainee driver. They were not a happy bunch and couldn't understand why I didn't have a bus for them. We couldn't find any private transport and ended up on a truck to Quetta. The train to Lahore was no better. It was stinking hot, 40°C in June. The border remained firmly shut, with all the different overland groups now congregating in Lahore.

By now I was due to be back in Delhi to take a group to Kashmir. Siân was also due to arrive in Delhi. A specially chartered plane was arranged to take all the overland groups from Lahore to Delhi. But to make matters worse, my group didn't have Indian visas and couldn't take the flight. In the end they had to go to hundreds of miles south to Karachi to get their Indian visas, and then fly to Delhi. My old mate Steve Merchant took them on to Kathmandu with some Dragoman clients who'd been stuck in Lahore but had managed to obtain Indian visas. I flew to Delhi business class, the only time ever.

I was back at the airport in the middle of the night to meet Siân, what chaos, but I then had to fly up to Kathmandu to collect the bus and drive back to Delhi as fast as possible. Not a good start…

4 August 1984: Siân

It was August 4th, my parents' wedding anniversary, and they had come down to London to see me off until some uncertain point in the future. My house was thoroughly cleaned for the tenants who would live in it while I was away, my hair was permed and coloured, and I was ready for anything. A day before I was due to leave, I received a telex from Bob in Lahore. 'Stuck in Lahore. May not be in Delhi when you arrive. Maybe you should wait...' But I already had everything arranged, and set off for the YMCA in New Delhi anyway...

At 2am the plane touched down on the hot Indian tarmac, and I was there. But would he be there? Some anxious moments followed, as I fought my way through the bureaucracy of the airport, and finally out through the immigration. Somehow he had made it, what a relief...

1984: Indian pastimes – Siân

Days turned into nights, nights turned into days, as I got over the effects of jet lag and slowly acclimatised to the sweaty midsummer heat. Whatever else one might say about Delhi, I couldn't say it was boring – a cacophony of sounds continually assaulting my ears, perpetual motion in the shape of runaway rickshaws, street sellers and beggars. The sweltering heat takes the urgency out of everything and I can now really appreciate the fact that I am in no hurry, free of the responsibility of a house, mortgage and job. It's definitely a city of considerable contrasts – from the dusty poverty of the backstreets to the self-indulgent luxury of the many international hotels.

The Golden Temple, Amritsar in peaceful days

In June 1984 the Punjab was closed to foreigners because of the trouble with the Sikhs – I went to the Home Ministry to try to get a permit for our private Hann Overland bus to travel through the Punjab to get to Kashmir. No way, said Mr. Khaka, 'If they want to go to Kashmir, let them fly there ... and if they have no money to fly, what are they doing here? I would like to be in Los Angeles at the Olympics, but I have no money, so I am not. If they want to travel by bus, take them south, there are many other interesting places to see in India. But we are not issuing permits for private buses to cross Punjab.' So there. We drove up through Punjab anyway and the border was unmanned – but 'that of course is India'.

Kashmir and the romantic houseboats of Dal Lake – 'old wooden sheds floating on a sewer,' said one client – and the unromantic corruption behind the scenes, petty arguments between the traders and houseboys, etc. etc. The boats are indeed quite luxurious – they do have flush toilets, which flush straight into the lake, along with all the other refuse and filth. Of course the water for tea comes straight from the lake, but, again, 'this is India'.

Beyond the 'fragrance' of Dal Lake is Ladakh and its barren, dusty landscape; friendly Tibetans industriously running restaurants and shops. Many historic Buddhist monasteries perch staggeringly high above stark rocky hillsides. The landscape is an incredible and luminescent mountainous desert. And how difficult it was to reach Ladakh! Our initial attempt was to fly to Leh with the first tourist group – but the flight never came. Only the waiting taxi drivers would tell us that the plane had been hijacked by Canadian Sikhs (before it reached us) and taken to Dubai.

We had no choice but to take our decrepit old English bus up the infamous Srinagar–Leh road; all public transport was fully booked a week in advance. Slowly we zigzagged up the steep, winding dirt track over the 4000m Zoji La pass, and down to the tiny village of Dras, where the Rest House had three rooms for our group of twelve people – we slept in the bus.

The next day we continued after breakfast, had no lunch, and found ourselves stopped in our tracks by a landslide that evening. We were in a military camp with no food or shops and one room barely big enough for ten people on the floor – so four people slept in the bus again! We did manage to get hold of some awful rice and eggs, which we cooked in the kettle – this staved off starvation but the taste was unmentionable! The following day, after a sleepless night, we waited and waited for the road to be cleared. By 7pm it was, so after another day without food we were back on the road. Slithering along a slimy road in the semidarkness, we were soon stopped again by another landslide on the muddy zigzags of the Fatu La and a traffic jam of crazily overloaded trucks.

This time there was no escape, no turning round and no source of food or water for miles. So the bus became the hotel for the whole group and Bob and I put up the one tent on the verge. On the verge of exhaustion next morning, we drove forwards up the hill only to find a heavy truck stuck in the middle of the 'road' – two feet of liquid red mud – and slithering towards the drop every time it tried to move. Dejected and full of apprehension, Bob reversed down the slippery slope, with an inch to spare on each side, for at least a mile, before he could turn round and start the journey back to Srinagar. Even this was not to be completed without problems. In mid-afternoon the engine blew, so we spent another night in the bus/local's house/animal shed and hitched a lift for twelve people in the back of a truck the next morning. What a story!

That was the end of their 'holiday' for the first group, but Bob and I had to return to the bus, fix the engine/gearbox/clutch at 4000m and get back to the next group.

Our next attempt to get to Leh was more successful – the most fantastic flight I have ever been on. With spectacular mountains all around us, the Karakoram, the Himalaya, we sped in minutes from the alpine pastures of Kashmir to the harsh desert scenery of Ladakh, with snow-white peaks towering high all around us. Dashing from side to side of the plane, not knowing where to look next, it was all so magnificent. Soon the mountains were above us and we whizzed narrowly past Spitok Monastery to land at Leh. So many travellers go to Leh to escape from India and it's easy to see why – cool fresh air, friendly people but no hassles, and the most incredible mountain scenery and monasteries.

Soon we were back in Srinagar and began the long drive south to Delhi, alone, through the green mountains towards the Punjab. We passed road signs saying: -

---

'This is not a race or rally, drive safe and enjoy Kashmir Valley'

'If married to speed, divorce her'

'Better late than the late'

Groups of nomads in rags herded their sheep from nowhere to nowhere and I wondered what drove them on. Down on the flat plains of the Punjab, a police patrol stopped us, searched the bus and warned us not to stop – the previous day a public bus had been stopped and five Sikhs had been dragged off and murdered. On that happy note we drove on for twelve hours until we were safely out of Punjab and into a comfortable motel in Karnal, where we shared our room with a mouse.

Delhi to Kathmandu

Back in the metropolis of Delhi, I found my birthday cards and we celebrated in Nirula's with a long-awaited bacon pizza and chocolate fudge sundae. In the swarming streets of Old Delhi, searching for spare parts among smells of sweat and spices; we huddled in a three-wheeler motorbike taxi surrounded by the humid, polluted congestion of the city.

Jaipur was our first stop on the route to Kathmandu. Just outside is Amber, the ancient capital of Rajasthan, with its castle, elaborate marble work, views from the hilltop and elephant rides 'like a Maharajah'. Jaipur, the Pink City, painted pink and rather shabby in places, bustles with crowds of people and small noisy vehicles. The Hawa Mahal or Palace of Winds, a pink facade for ladies to hide behind while they watched the proceedings in the street, looks resplendent in the early morning sunshine. The Janta Manta Observatory brings back memories of elementary physics lessons, as you try to puzzle out how things work, and I climbed to the top of a huge sundial for a superb view over the town. The City Palace, another splendid building, houses amongst other more pleasant things, a horrendous collection of weapons used for the most unspeakable purposes. Khetri House, our hotel in Jaipur, was a Maharajah's palace, but the Maharajah now lives in America with his girlfriend for most of the year. (Sadly some years later it closed after a family quarrel and never reopened.) Mr. Pram, our tourist guide, gave his usual lectures on the history of the area, the dowry system, the North Indian dacoits, and a few amusing remarks about Mrs. Gandhi. He then proudly invited us back to his home to celebrate his birthday with a glass or two of dubious Indian whisky!

Eugene Pram, the colourful guide of Jaipur

On the road again, we crossed the great North Indian Plains to Agra. This is a long, straight flattish road, with stately camels pulling heavily-laden hay-carts, and the occasional sad-looking bear on a rope paraded out into the road for the tourists. Very thin cows that look as if they've come straight out of an anatomy lesson might choose to move lethargically out of the way of our 40kph speeding bus, if we're lucky.

Before Agra lies the fascinating deserted city of Akbar, Fatehpur Sikri. This ancient red stone Moghul city with its white mosque is very well preserved, and is a beautiful place to wander round, even in the sweltering heat which seems to descend on it at all times of the day. Unfortunately the children at Fatehpur Sikri are some of the most objectionable in India, continually pestering the tourists and trying out the latest English swear words, but they cannot detract from the spectacular architecture. From the Red Fort in Agra, the Taj Mahal can be seen gleaming in the distance across the river. On my first visit, the monsoon rains started to descend as we gazed at the view – soon we were forced to wade through inches of water to return to the bus and drive to the splendour of the Taj Mahal itself. Glistening in the rain, the paths around it shining brightly, masses of large black umbrellas in the foreground, the Taj looked more impressive to me on this day than on my second visit in harsh sunlight.

It's a long day's drive from Agra via Gwalior to the erotic temples of Khajuraho, the first five hours non-stop to the busy town of Jhansi, where a loo-stop was urgently required. This filthy town did not appear to have such a thing as a ladies' toilet, so I found the three-walled concrete structure, which was the gents', and dashed in there cross-legged. We had dhal and chapattis for lunch in a rather seedy roadside café which Bob had visited several times before, with no ill effects. Back at the bus, we had to forcibly part the crowds that had gathered, in order to get through the door.

Onwards through country towns and along narrow lanes, with a 40km diversion because the road had been washed away by floodwater. We eventually reached Khajuraho in the late afternoon. A power cut delayed our dinner at the Raja Café, run by a Swiss lady and her two sausage dogs. The dogs looked like they'd eaten too many left-overs at the café, but the one they were taking for a walk looked immaculate in her finest sari! In fact there were two sisters who ran the place; both lived out in the countryside with their unseen husbands running farms. Perhaps they were the last vestiges of the pomp and circumstance of the British Raj.

The impressive fort at Gwalior, India

In the morning we wandered round the temple complex, wondering how one could perform the complex acrobatics depicted in the sculptures. As it says in the guidebooks, the characters would win a few prizes in the sexual Olympics! The temples are set in a lovely park, with neatly-cut green lawns and bright red flowers beneath a clear deep blue sky – very relaxing and peaceful.

Saucy artwork at Khajuraho

Satna, on the way to Varanasi, is a nondescript Indian town with a nondescript hotel and nondescript food. Our toilet reeked, first of stale toilets then of strong disinfectant when it had been cleaned with a cursory wipe of a mop. Next morning we were on the road to Varanasi, with no stop for lunch or even a cold drink all day. There is nowhere to buy anything and the locals have been known to throw stones at buses.

What a relief to get to Varanasi, and the Dak Bungalow with its large rooms, big but hard beds, and cold showers. And cold drinks, since we had not quenched our thirst all day. The bus was taken straight to Shafiq's yard – within minutes all four wheels were off to investigate the brakes, and the gearbox was off to try to determine why the clutch kept failing, etc. They are such characters at the Dak Bungalow – the old man who's always going off duty and won't be back and can he have his tip now? – the masseur 'baby feet, baby feet, very good' feeling everyone's feet for the lumps of hard skin – the manager who wants everyone to eat in his restaurant and buy his silk – and of course Shafiq, the manager of the garage across the road that mends all the buses or at least keeps them going until the next visit!

Varanasi is the holy city of India and every Hindu wants to go there to die, so that he/she goes straight to heaven. This seems to result in a 'don't care' attitude to everything. The town is filthy; people drink water direct from the holy river, the Ganges, where all the dead bodies are thrown (and the sick bodies are not even burnt). Pedestrians have a suicidal way of walking into vehicles, which is not encouraging to a European driver who will get slung into a long-forgotten prison if he hits someone.

The local city tour begins at 5.30am when one should still be fast asleep, but we dragged ourselves awake and were taken down to the river for a cruise along the waterfront. First we saw the holy pilgrims bathing in the filthy water, then we saw the bodies being cremated and the funeral ceremonies – enough to wake anyone up by now. 'Did you know that it has been scientifically proven that the water of the Ganges is 100% pure? It is difficult to believe but it is true,' said our guide. I certainly found it very difficult to believe, especially when the city's sewage joins the diseased bodies in the river.

Local philosophers, Varanasi

After the boat ride we were taken to the Mother India Temple, which I found particularly fascinating for its large relief map of Asia and the Tibetan plateau. At this time we had no idea we would be in Tibet six months later, but I was amazed at the vast expanse of the plateau, its height above India and its flatness. Finally our tour headed to the Monkey Temple with its huge families of monkeys and worshippers. A day later a tourist was bitten by a monkey and had to undergo a two-week series of anti-rabies injections at the local hospital, but we emerged unscathed.

Varanasi ghats on the Ganges

Hann bus & bullock carts, which transport is faster?

And so to Gorakhpur, the last town in India; a crowded bustling city whose prime attraction is Bobi's Restaurant. In this dimly-lit restaurant with a certain subtle ambience, our group of sixteen ordered, were impeccably served, ate and paid for their meals in less than an hour – quite an achievement for any restaurant in any country. So what if most of us did have our usual meal of vegetable cutlet, chips and coffee?

Continuing up what could be the worst main road in the world – three hours to cover sixty miles – passing the odd elephant on the way, we all felt a great sense of relief at being in Nepal. Having been to Nepal only once before in 1982, I had been longing to go back ever since and I was not disappointed. Even the scrummage at the border was somehow different from the other side, and members of the group who had not been there before also noticed it. However, things were not to be so simple yet. Royal Nepal Airlines had cancelled a flight and put their passengers in our hotel, so we had no room, only a skimpy omelette for dinner, a bus limping along with a broken spring and an exhausted driver. But on we had to go – thirty miles to the next town, where we crammed into a Tibetan hotel, three to a room and two of us in the bus.

The morning was clear as we dragged ourselves up from the floor of the bus and drove back into town to pick up the group. After breakfast we really felt we were in Nepal as we drove along the most magnificent mountain road: green terraced hillsides and the snow-capped Himalaya in the distance. The winding road took its toll on Bob's arms though – seven hours of continually turning the steering wheel in different directions left him barely able to move his shoulders.

Of course when we arrived at Pokhara the hotel was fully booked and they had not received our telegram, but we were used to this by now and anyway the manager's brother had rooms in his hotel! They did have food, so months of vegetarianism in India were broken by a feast of pepper steak – succulent buffalo steak tasting as though it had been marinated in red wine, and apple pie – oh what luxury! Pokhara is a sleepy sort of place, situated on a lakeside and dwarfed by the peaks of the Annapurnas, Machhapuchhre and Dhaulagiri towering high above. The lakeside town is full of restaurants where you can treat yourself to cheese omelettes or banana pancakes while sitting on a rooftop gazing at the Himalaya. Various shops sell anything from heavy bread and cheese for trekking to the usual tourist handicrafts. Occasionally a character staggers across the road totally stoned, but I didn't see much evidence of drugs.

Siân and the Hann bus at the Kathmandu Guest House 1985

So to Kathmandu, to where I had long dreamed of returning; a mediaeval city full of character and life. Some sleepy side-streets; others so bustling you can't walk in a straight line. Colourful handicrafts hang everywhere; brightly-coloured sweaters, multi-coloured Tibetan carpets made by the refugees, scarves, gloves, jewellery, leather jackets; you name it, they have it, and more often than not, it works. We stayed at the Kathmandu Guest House.

Old alleys in Kathmandu (above) and Bhaktapur (opposite)

Kathmandu is a great mixture of old and new; the Nepalese have so far succeeded in creating new buildings without destroying the character of the old town. Few tourists would complain that restaurants exist where they can eat pizza, Spaghetti Bolognese or apple pie after weeks on trek on a constant diet of runny dhal and rice!

Bhaktapur Durbar Square

Soon we had to leave this paradise (after customs formalities regarding the overstayed bus had been sorted out) and return empty to Delhi to collect the next group. Having finally crossed the border into India, we were first stopped by a puncture. That mended, heavy rain started to fall and since our windscreen wipers would not work we spent two hours by the roadside listening to the Dire Straits we seemed to be in. The rain stopped, we carried on, and a second puncture halted us in our tracks. As we didn't have two spare wheels, Bob struggled with his inadequate tools in the increasing darkness, and eventually prised the burst tyre off the rusty old wheel, with the help of a few passing Sikhs. So on we continued with only 'five wheels on our wagon'. At midnight we were lucky to find a quiet lay-by, where we fell asleep in the back of the bus. But it was not quiet enough and, after a few restless hours wondering if the next truck would come hurtling into the back of us, we were forced to move on to another 'quiet' spot.

By 6am we were on the road again, for what was to be an extremely long day. On and on and on, through Faizabad, then to Lucknow. Here we had a quick look at the Residency where the British were besieged. At 9pm exhaustion overtook us, so we started to prepare a meal of Heinz macaroni cheese, in a side road, which we took to be deserted. But nowhere in India is deserted, and within minutes a man appeared and told us 'danger', so we gulped down our luxury meal and Bob decided to continue to Delhi. It would be meaningless to quote distances; suffice to say we drove constantly till 2am the next morning and then lay on the floor of the bus again until breakfast in the YMCA at 8am.

I heard later that the group were more than a little surprised to see us that evening unwashed and wearing the same clothes we had had breakfast in. All day we rushed around Delhi, collecting mail from the GPO, getting money from Thomas Cook, checking for telexes, an excursion into old Delhi for more spare parts for the old wreck, and finally a few minutes to relax over bacon pizza and chocolate fudge sundae. Then the pre-departure meeting, at which we found a very nice group, staggered by the story of our journey.

On schedule, to Bob's relief and delight, we left the next morning for Jaipur. Along the flat 'interstate highway', past the camel-carts and bears, stopping at the Midway restaurant for the usual vegetable cutlet and coffee, and soon to Amber; but only just, for the engine started to emit the most terrible noises.

This was where my induction into group management began, for from here on I had to take the group all the way across India on public transport. Bob limped into Jaipur with the bus while the group were visiting Amber, and I waited for them with the news that there was something wrong with the bus and Bob had taken it into Jaipur to start work. Which of course was true, and he didn't know the extent of the damage then. We ended up on a public bus to the busy town centre; there we all piled into one 3-wheeler trishaw to rattle along to Khetri House. A disgruntled and depressed Bob greeted us, with the engine already in pieces and not looking very hopeful.

To cut a long story short, we continued the trip by public deluxe bus and train to Varanasi. In Agra I met a tour operator who was just starting in business and he gave me a very good deal for taking the group to see all the sights in his minibus. We spent a couple of afternoons lounging by the pool in Clarks Hotel, eating in their superb restaurant, all free of charge for the 'group leader' – and I got the manager's card and a request to the Varanasi Clarks hotel to give us the same 20% group discount.

The overnight train to Varanasi was not so much fun. I found it strange that some of the group had already been on Indian trains and I, their leader, had not, but that was no problem – however, one of the Irish girls had her passport stolen in the crush and had to return to Delhi for a replacement. Otherwise all arrived safely at the Dak Bungalow in Varanasi, with its plentiful cold Limca and chips. From there I hired a bus to take the group to the Nepalese border and another driver Hugh was telexed to collect them there in his Hann bus. I flew back to Agra and waited for Bob with the new engine.

Now we were driving to the Nepalese border ourselves to pick up the next group one week later, so we decided to stop in Lucknow to see the sights properly. It was well worth it – the Great Imambara, many mosques, mazes and rounded domes all over the city, with excellent panoramic views. But we couldn't eat in our favourite Kwality Restaurant; a sad but aggressive group sat beneath some posters on the front door – perhaps someone had been killed here? Next it was on to the infamous Gorakhpur with Bobi's restaurant and the Ambar Hotel. On our previous visit a mob had thrown stones, which cracked the back window, but this time it was more peaceful.

Another trip, another breakdown

> **Mrs Gandhi is assassinated**
> Sitting with our vegetable curry and chips the next evening on the Nepalese side of the border, we were quite stunned by the waiter's broken English: 'Mrs. Gandhi dead shot'. We could only look at the pictures in the newspapers and hear odd bits of news as they trickled through via the hotel manager. (Mrs. Gandhi had been shot dead by her own Sikh bodyguard.) Then our next group arrived – 26 to go on a 25-seater bus – and they had first-hand accounts of the violence. They had been sightseeing and shopping in Varanasi during the afternoon, and within minutes of the shooting being reported their taxi-drivers came searching for them. Shops were hurriedly boarded up, Sikhs disappeared into hiding and the looting and fighting began.

The next morning they left Varanasi as quickly and unobtrusively as possible, in their smart Indian bus with Sikh driver. They heard reports of tourists being offered safe conduct out of Varanasi by the army, and then having all their films confiscated; fortunately this did not happen to them and they reached the haven city of Gorakhpur without incident. No stopping for lunch though; all the restaurants were closed, shops were boarded up, the streets were empty – it was almost a ghost town. At the next insignificant town, trouble arose. Two buses were set on fire in front of them, tyres burning, black smoke billowing forth and all the windows smashed. Then the mob turned to them, seeing the poor Sikh driver in the cab. Before anything nasty could happen, the group piled into the cab with the driver and the show of white faces fortunately dispersed the mob. They continued to the border a little shaken but without further incident.

Even the rest of the journey to Kathmandu was not entirely trouble-free, as we developed a leaky oil pipe. That fixed temporarily, we struggled uphill with our heavy load and me sitting on the toolbox in the gangway. A welcome sight was Hugh coming the other way, on his way alone to Bombay, very apprehensive indeed about entering India in such a riotous state on his own. We all wondered what had happened to the Sikh driver the group had left behind at the border – was he in hiding at Sonaull, or was he trying to return to Delhi? Hugh talked to the group for a few minutes and instantly decided to return to Kathmandu, so we split the group and I finally got a seat. Back in Kathmandu we found that the border was in fact closed for several days while the situation in India calmed down.

Interlude in Langtang: November 1984

A week later it was time for a holiday and my sister Kathy flew out for her first trek in the Himalaya. I of course had my first really knockout dose of sickness & diarrhoea; two days before the trek I was lying mindlessly in bed, dispersing liquid from both ends with great force and frequency. However, with the attention of my own private doctor I was fit enough in time to stagger off to the 6am bus to Trisuli, where we started our trek up the Langtang Valley.

The first night found us in the tiny village of Bogota, in an empty house belonging to a Brahmin family who were housing the Peace Corps worker who taught in the village school. I believed the empty house had been occupied by a young couple; the husband had died through illness and his widow had become pregnant and remarried within less than one year. For this crime she had been banished, penniless, from the village and sent to wander the mountains with her new husband. Her previous children were kept by relatives in the village. It was a rare privilege for us to be allowed to eat with a Brahmin family – we were treated to dhal, rice, potatoes and cheesy yoghurt, and we were even offered spoons in case we did not want to use our fingers in the traditional Nepalese way. First of all we sat cross-legged on the floor to eat; when we had finished the family took their places and we left the room. The children were obviously delighted to see us and sample the mince pies Kathy had brought from England, and we were happy to have a relatively comfortable and warm night's sleep.

The next day we left early, stopping for tea at a roadside house, and continued along a pleasant path with superb views. Occasionally I started to flag a little, having not eaten a proper meal for four days, but the odd tin of pork & mushroom paté soon revived my spirits and energy. At 4.30pm we were passing a series of road-workers' shacks and a lady invited us to stay for the night, so we stopped and lay out our sleeping bags on the rough wooden platforms. Dinner was the usual dahl and rice, but with very tasty potatoes and beans also. By 6pm Langtang was clearly visible in the golden evening light, but it was soon pitch black and icy cold so we retired to our sleeping bags to get warm. The family curled up on the opposite mattress, having first padlocked the wire netting walls around the 'house'. I just hoped I wouldn't need to get up to irrigate the ground during the night!

The trail continued along the road to Dhunche; a road sometimes blasted away so that huge boulders would hurtle downwards on to the original path, trampling anything or anyone that got in their way. In some places skinny men in rags would chip away with tiny chisels at enormous chunks of rock, a hopeless and thankless task.

After Dhunche the inevitable ups and downs began. Down to the river, up the steep rocky hillside opposite, along a bit, down again, and so on. We reached Syabru at 5pm, as darkness fell, and relaxed in the first lodge we stumbled into. It was a charming little place; one large room with a long mattress along one side, the family's bed in one corner and the kitchen in the remaining corner. The kitchen shelves were full of lemon cordial, beer and spicy sauce for cooking; the open fire in the middle of the room warmed us as we huddled around it; the lady of the house was breast-feeding her smiling first-born son as she prepared our French onion soup, noodles with vegetables, and 'pancake jam'. Out in the middle of nowhere, I could hardly believe my eyes when I saw the extensive menu; it was amazing what she could make from such limited raw materials.

Down to the river again, and up the next day; down through lush green forest to a clearing by a rushing stream where a few other trekkers were eating lunch and sunbathing. We continued to the main river before stopping for our lunch of duck & orange pate followed by Cadbury's chocolate. Mmmm! Just as we started to rise from our comfortable slab of rock, a bedraggled and breathless figure emerged from the forest.

Bob had caught up with us after leaving Kathmandu one day later. Kathy was now free to zoom on ahead. Having been an Oxford Blue rower, she was much faster than me, so we met her at the wooden Riverside 'Hotel' drinking copious cups of lemon tea that evening.

From this point on the trek diary is blank, so I write from memory some 30 years later!

Siân and Kathy above Kyanjin Gompa, Langtang

In those days there were scarcely any lodges, and certainly no single or double rooms. Trekkers lined up their sleeping bags on wooden platforms around the central cooking fires in the smoky living rooms. Menus were generally simple and very limited. However, at Kyanjin Gompa there was a new, almost Swiss-style lodge. In fact the Swiss had a project here making cheese from yak (nak) milk. Fresh cheese and yoghurt was a real treat. Beyond here the trail heads towards the Langshisa Glacier, where the altitude is above 4500m. Breathtaking views surrounded us.

Bob high above Kyanjin Gompa, Langtang Valley

Not long after the trek it was Christmas and New Year and there was no more work.

Lhasa to Kashgar: February 1985

We left Delhi and flew to Thailand, spent a week or so recovering on Ko Samui and then continued on to Hong Kong and the Philippines. Back in Hong Kong, student cards were bought from a dubious fellow in the stairway of the infamous Chungking Mansions in Kowloon. How this building passed any fire regulations is beyond belief, except that a lot of people were making hot, dodgy money within its confines! Our Chinese visas were issued; we were on our way to Tibet.

The eastern ranges of the Himalaya sparkled as a pink sunrise revealed a magnificent panorama of white peaks. To the south we spied Namche Barwa, a mountain cloaked in mist and mystery. To the west the great lake of Yamdrok was just visible as the plane lurched and began its descent to Gongkar airport. Like hyperactive children, we both made for the exit of the plane, to be taken aback by the rarefied icy blast of the Tibetan plateau. Rolling brown desert-like hills surrounded the airport. The luggage could not be collected, although no one knew why. The road was as abominable as could be imagined. Snow men were sweeping off the light dusting of fresh flakes along the road as we climbed.

Lhasa, the Forbidden City, is a jewel. Since time began, so many explorers have perished or failed in their quest to enter its gates. Our bus drove into town and parked in the grubby Chinese compound of the airline below the Potala Palace. The luggage would not be available for collection today. Somehow this was not the way it was supposed to be. The Chinese in Lhasa were not a happy bunch, stuck in this harsh, uncompromising land, far away from their green homelands. The Potala defied description; a fantastic fairytale building that in reality was so much more overwhelming than any of our preconceived notions. Could there ever be such a building anywhere else but Tibet?

Over the Eastern Himalaya

Nearly all the travellers stayed at the Banak Shol Hotel not far from the Tibetan heart of Lhasa. Tibet had only just opened to independent tourists, and the Snowlands was another new place that opened the day we arrived. This meant that it had super-clean new bedding and thermos flasks that had not yet been dented. We ate yak burgers at the Banak Shol; everyone was full of excitement about the travel possibilities in Tibet. Lots of aspirin were consumed; Lhasa is unkind to those used to living at or just above sea level.

Potala Palace, Lhasa 1985

Recently a few tourists had been allowed to visit Xigatse, the seat of the Panchen Lama and the location of the Tashi Lhunpo Monastery. Some had seen Gyangtse and a few had ventured west. Unfortunately some indiscreet and inconsiderate tourists had climbed over the forbidden walls in their attempts to visit parts of the Tashi Lhunpo that were restricted. As a result of this, since then all others had been forbidden to travel outside Lhasa. Of course we were all frustrated at not being able to go into the countryside, but these were early days for the Chinese with regard to their liberalising tourism and travel in Tibet. In retrospect it was very fortunate that we had the opportunity to visit Tibet at all, as individual travel became impossible after riots and demonstrations by Tibetans demanding freedom from Chinese rule a couple of years later.

There was a wealth of places to visit in Lhasa. The two great monasteries of Drepung and Sera were accessible. In the city the Norbu Linka, the Dalai Lama's summer residence, was open, the place where as a boy he had been taught and entertained by the German climber and war escapee, Heinrich Harrer. The central Barkhor bazaar was a lively place with many country people. Khampas from the east were dressed in amazing yak-skin coats and sported long hair, some with plaits and red decorations. Mountains of butter for lamps and cooking were stacked up in one corner; pilgrims circulated the Barkhor and prayed before the holiest chapel in Lhasa, the Jokhang. Sometimes a Chinese resident would cycle around the Barkhor in the wrong direction, anti-clockwise, just to upset the Tibetans.

Even as outsiders it was possible to experience the everyday tensions and frustrations of the Tibetans. A thinly veiled sadness pervaded the atmosphere. We visited the old medical hill, the Chokpuri, and climbed the long steps to enter the Potala. Boisterous and excited country people accompanied us into the first of the great courtyards. The whole building was built in stunning proportions. Hundreds of windows with Tibetan-style black-painted surrounding borders pocked the outer wall of white and red. The colours contrasted strongly with a deep blue, almost iridescent sky. On the top floor were the living quarters of the Dalai Lama, in a yellow palace with golden deities and many prayer flags.

From the rooftop the views of Lhasa were literally breathtaking. In one amusing section we were able to see the toilets; the drop under the holes was staggering, disappearing into the dark abyss far below. I remembered my first experience of French squat toilets when I was a teenager, and fifteen francs of my precious pocket money had fallen into the hole – this would be a far worse place to drop any valuables!

Backstreets of old Lhasa 1985

Barkhor 'belles', Old Lhasa

Khampas, Old Lhasa

Kids play around Drepung Monastery

In the chamber of the demons and gods, 20ft-high statues cast eerie shadows. Barely lit by flickering butter lamps, these grotesque characters with contorted faces seemed to be watching us. Their eyes followed us as we passed by. Others were demonic in appearance; some posed like the equally sinister-looking Black and White Bhairab images of Kathmandu's Durbar Square. The major icons were also on display, the Buddhist deities, Guru Rinpoche, Chenresig, Avalokiteshvara and Sakyamuni Buddha. The old Bonpo animistic beliefs, Buddhism from India and the somewhat esoteric Tantric practices within Tibetan culture have all combined to form a uniquely strange and mysterious form of religion in Tibet. Much of this mystery is the draw that pulls many people to become fascinated with the Tibetan culture. The mountains are another, the pure air of the plateau and the many spectacular monasteries add to the air of mysticism. We felt this captivating spirit, perhaps a dreamer's nostalgia for the Tibet of old. It is a feeling not dissimilar to that felt in the presence of the great stupas of the Kathmandu valley. We were saddened and captivated simultaneously by Lhasa and Tibet.

7 February 1985: Lhasa

Little did we know that this was to be our last day in this alternately fascinating and frustrating city. Jonathan, our crazy English friend from Guangzhou, and I went off to the Qinghai bus station to buy tickets to Golmud, leaving in three days' time. At least, we asked for tickets to Dun Huang, an open city famous for its cave paintings and on our way to the far west of Xinjiang. We 'didn't understand' when we were only sold tickets as far as Golmud, two-thirds of the way to Dun Huang, but a very closed city. Having had no success in getting out of Lhasa towards the Himalaya legally, we were quite determined to see something of the Tibetan countryside by travelling overland. On our return to the hotel, we found that our proposed hired Land Rover trip to Xigatse was cancelled due to lack of money and trust in the Chinese – would we or would we not get fined by Public Security for organising this? So we decided to get out of Lhasa while the going was good and I returned to change the tickets and leave tomorrow. Everything then became a bit of a rush.

We returned to the market for a last look round, to take more photographs, and for me to buy some Tibetan boots. These are worn by most of the Tibetans in the market; they are made of thick dark wool/felt with bright red and green embroidery and thick cord soles, presumably sufficiently warm and hard-wearing to protect their wearer against the bitter winter cold. With these tucked under my arm I continued to wander round the busy streets; the Tibetans were obviously delighted that I liked their boots, as they kept coming up to me and touching them, patting my shoulder and smiling broadly. The Chinese were noticeably diffident and ignored this as they walked the wrong way round the Tibetans' holy temple in the middle of the market.

Potala from Chokpuri hill 1985

I then chanced upon an American couple who were going up on to the rooftops to visit a boot factory, so I joined them for this mysterious excursion up a series of rickety staircases. Eventually we emerged on to the roof and the factory – a small attic shed where five women sat, hand-sewing the top of the boots and attaching it to the thick soles with coarse thread. The American was trying to negotiate a deal to export boots to Dale, Colorado as après-ski wear – I wondered if he would be successful or if the Chinese would step in and make it financially unprofitable.

After a few more moments to gaze at the magnificent Potala across the rooftops, it was time to return to the hotel for our last meal of Mapu Dofu. And so to bed early before our arduous ride into the unknown.

8 February
At 6:15am we were ready to leave but the hotel was locked. So much for our story about moving to another hotel. We woke the Tibetan lady in charge and she smiled as she let us out into the dark. Walking down the road in the pitch blackness, I felt like an escaping criminal. Had we been tricked? Would Public Security be there to meet us at the bus station, or would our great adventure succeed? At least we were unlikely to be disembowelled and decapitated if caught like the 19th-century 'trespassers'. We seemed to be in luck, but the bus looked appalling. It was an ordinary local bus, with upright plastic seats, no headrests, no heating and several broken windows. Dawn was breaking rapidly as we sat anxiously waiting for the bus to leave, but the only people to arrive were three more Western tourists, also looking apprehensive. Soon the bus was full enough to depart, overcrowded with Tibetans in sheepskins sitting on the steps and between the seats... and the prospect of a thirty-five hour non-stop journey over 18,000ft (5500m) passes through the icy cold of the night.

Sitting bolt upright, wrapped in all my clothes and my sleeping bag, we jolted off into the semi-darkness; past the sleeping check-posts and out of Lhasa into the wilderness desert of northern Tibet. As dawn broke we gazed out at small villages, massive herds of yaks, sheep and barren landscapes with the towering icy peaks of the Nyenchentangla in the near distance. Occasionally herdsmen rode past in brightly-bound sheepskin coats; despite my warm layers I shivered to look at them.

Nyenchentangla range north of Lhasa 1985

Our lunch stop was a highly dubious café, only the steamed bread seemed worth the risk – then on and on we continued. The scenery became flatter but still quite barren and desolate... we had a loo stop with not a boulder in sight, but by this time my main concern was not whether anyone would see me, but whether I would freeze solid if I lowered my trousers. So we continued, some brown rounded hills; this may be good grazing in summer, but where can the winter food come from? Dinner was fried vegetables with hot spices; there was no toilet, we simply had to wander round behind the restaurant and hope we didn't tread in anything too unsavoury.

Then came a night so cold, uncomfortable and interminable I can hardly describe it. Jonathan apparently thought he was going to die at one point – he had no sleeping bag and insufficient clothing for the -30°C temperature, so he tried walking up and down the bus to generate some heat. Most of the Tibetans had got off at the last village so we had room to spread out a bit. I lay across the double seat behind the door, but was in danger of falling into the stairwell as we lurched along the road. If I took my hands out of my bag to hold on, I was too cold. I tried lying on the floor on my Karrimat but the vibration was too much...

9 February
I looked at my watch. It was about 2.30am and we were passing through the eerily moonlit snow-covered sand dunes of the Tang La pass. Despite the bitter cold and lack of sleep, I thought this was one of the most incredibly beautiful sights I have seen. So peaceful and unspoilt, we certainly seemed to be trespassing on the roof of the world here.

Daylight came at 9am and we drove on and on with not even a loo stop. But there was nowhere to eat or drink either, so it didn't really matter. We passed through more snow-capped hills as we approached the truly forbidden city of Golmud. Dumped in the middle of nowhere at 3pm, we staggered in deep exhaustion to the Public Security office where a friendly cup of tea awaited us.

Surprise, surprise – Public Security told us there was no bus to Dun Huang and we would have to hitch-hike. They pointed in the direction of the hotel and after another long walk we arrived there. In this supposedly closed town, the hotel only takes foreign exchange money. Anyway, my main priority was to have a drink, a rest and a wash. Seeing the relatively clean toilets with running water, I thought we might be in luck and asked where the showers were. 'Mayo (No)!' was the reply as usual. I managed to sneak a wash in the sink while Bob and Jonathan went off in search of the truck station to try to negotiate a lift to Dun Huang tomorrow. By the time they returned the restaurant was shut, so we were stuck with tinned luncheon meat (Wu Can Rou) and soggy bread.

10 February
Up at 5:45am, but for what? After breakfast of tinned pineapple and sliced pork rolls, we left our first centrally-heated bedroom in weeks. Out into the bitter cold darkness of the Golmud road, and a long walk to the hitching spot. After two to three hours of dancing in the dark to keep our toes alive, no trucks came, the town still seemed asleep, and an anonymous 'helper' taxied us back to the hotel we had left earlier. I stayed with the baggage while the lads went off in search of a bus station, railway station, truck station or indeed any way of getting north. Nothing seemed likely to happen till tomorrow, so we checked back into the hotel and I washed some clothes for the first time since our flight into Lhasa. In the afternoon we strolled around the 'sights' of this remote closed city. There was a fantastic choice of food in the shops – tinned kiwifruit, tomato ketchup, apricot jam; huge stereo systems, cosmetics in abundant supply, and the usual incredulous stares and whispers from the locals.

11 February
Again we staggered through the cold, this time to reach the railway station at 8am. 1½ hours later the train was due to leave, but we were still unable to buy tickets for the intermediate station we wanted to get off at to continue our journey to Dun Huang. It was suggested to us that we should return to the truck station and hitch to Dun Huang. However, no sooner had we sat down on the bus to town than we were hastily recalled and bundled on to the train. Where next, we wondered?

As the train pulled out of the station, we were told that we would have to go all the way to Xining, many miles to the east, not arriving until 1pm the next afternoon. No arguments would persuade them to let us off earlier, so we sat back, resigned to an unexpected 27hr journey. I had a heavy cold, which would probably have kept me indoors in England anyway, so I was not too unhappy at the prospect of doing nothing for a while. And we did by chance have six bread rolls and a tin of Wu

Can Rou to keep us going. As we rolled slowly through the 'countryside' – no more than 40mph I would guess – I could see why they didn't want to let us off. The railway line ran parallel to the road, a dirt track with barely any vehicles, running through barren desolate desert with no habitation.

Some Chinese people joined us at a later station, so we retreated to the top bunks for some sleep. These were quite comfortable and way above the floor; no danger from pickpockets and. a vast improvement on the Indian 2nd class sleeper.

Hard sleeper class

12 February
Rolling ever onwards through the desert plateau, with isolated hills here and there, we arrived at Xining at lunchtime. A helpful Beijing Chinese man bought our hard seat tickets to Urumqi, reducing our waiting time in the interminable queues at the station booking office. So we had some time to look around the old streets of Xining and stock up on Wu Can Rou, bread, chocolate and tinned kiwifruit for the 2½ day journey ahead. It was very, very dusty and windy – some old Uygur men sat by the roadside, a couple of unusual mosques were shaped like square Chinese temples. Wide modern Chinese streets surrounded picturesque narrow dusty stone alleyways.

To our later regret, we didn't go to the Taer Su Monastery out of town; mainly because of lack of time and enthusiasm with the thought of our long journey to Urumchi. We also thought it would be hard to beat the monasteries we had just seen in Lhasa. After some hot spicy kebabs outside the station, we crammed into the hard seat compartment of the train to Lanzhou; very crowded but we were lucky to have a seat at all and it was only for five hours.

Xining mosque 1985

Xining backstreet 1985

13 February
Before we could lie down on the station floor and get some sleep, we had the customary hassles with the railway staff. Having sorted that out, our evening meal (by now it was 1am) consisted of two bread rolls with Chinese tomato ketchup. Only then could I roll out my Karrimat and sleeping bag on the floor, as I had done ten years ago in Belgrade as a student on InterRail. Our next train arrived at 3.30am and we fortunately got a hard sleeper within half an hour, climbed up to the top again and fell asleep immediately.

At 6am we awoke to the rasping sound of Chinese propaganda and music, blaring forth near my left ear. Then a comedy program, which seemed to amuse those Chinese who could decipher the crackles and understand the jokes, but was not very funny to me at that time of day. We rolled slowly forwards through what seems to me the biggest expanse of desert and scrubby pasture in the world. Sharing our compartment were a middle-aged Chinese couple with the biggest mugs I've ever seen – at least two pints each. Wu Can Rou and ketchup rolls are starting to lose their appeal.

14 February
Onwards ever onwards... at 30–40mph to cover thousands of miles, no wonder it takes so long. The staff are always sweeping and mopping the floor and toilets, removing the spat-out sunflower seeds and phlegm that are constantly evacuated from Chinese throats. We think our carriage won a prize for 'cleanest carriage' since there was great applause and congratulations after a garbled announcement on the loudspeakers. Outside, the desert continued, broken occasionally by some craggy canyons and a few camels. For a change from Wu Can Rou, we tried lunch from the restaurant car; it was surprisingly good and tasty. Music continued to escape from the horribly distorted speakers, followed by more comedy programs with canned laughter that did not amuse even the Chinese passengers.

Finally our journey ended at Urumchi at 10.30pm, and we had to leave the cosy warmth of the train to venture into the icy cold of the night. Out into the deep-frozen snow, coaxing our legs into movement after nearly four days' stagnation on the train. Fortunately Bob had visited Urumchi before, so he took us on a long walk past a motley collection of late kebab sellers to the Xinjiang Overseas Chinese Hotel. Here the three of us were given a dormitory; a three-bedded room with comfortable beds, central heating, hot drinking water and showers – blissful dreams come true.

Urumchi panorama 1985

15 February
As usual, the first thing to do in a Chinese town is to buy your ticket out, and as usual, that was easier said than done. For some reason: 'No flight to Kashgar tomorrow, but maybe on Sunday – come back tomorrow.' We heard conflicting stories about the Spring Festival (Chinese New Year) travel arrangements. The bus company said they continued as normal throughout the three-day holiday, and so did the airline CAAC, but CITS said that everything would stop for three days. I suspected that CITS wanted a three-day holiday for themselves, so we decided to try CAAC again tomorrow rather than commit ourselves to a four-day icy bus journey after all those nights on the train.

It snowed heavily all day, so it was difficult to see anything and my hands froze up instantly when I took a picture of a mosque. The local buses were more than full to capacity and the roads were sheets of solid ice. I tried to get on one bus but was only partially successful; as I hung on with one hand and another girl behind me, the bus slithered off along the road. My hat fell off, but she caught it between her body and mine and I retrieved it at the next stop. Not wishing to risk falling off while the bus was moving, we got off here and walked the rest of the way to the bank. Urumchi appears a drab almost eastern European-style town with only a few older buildings left. Lorries skidded to a halt all over the road as we walked back to the hotel, stopping for a meal of kebabs and flat bread and warming our frozen fingers above the glowing charcoal. Back in our centrally-heated room, we certainly weren't going out again for dinner.

Curious kids look on

16 February
It was still very cold (approx. -30°C) but the sun was shining brightly, so we were able to appreciate the pleasures of Urumchi. Three hours of waiting in the CAAC office resulted in two tickets to Kashgar tomorrow, so we could relax. We spent a fascinating afternoon wandering around the old streets – mud buildings and squalid houses, muddy roads, mosques here and there. The temperature was just warm enough to take a photograph before my hands froze painfully solid. None of the guidebooks do justice to this town, saying it is simply a hideous industrial centre and not worth a visit – on the contrary, I found the old streets extremely interesting, though the temperature could be improved on. Jonathan had already left us to travel by bus to Kashgar, so we had hot showers and an early night, a luxury not to be repeated for a long while.

17 February: Kashgar
We finally made it to the exotic outpost of Kashgar after a rather different CAAC flight. Bitterly cold, we first waited by the roadside for the bus to the airline office, then the bus to the airport, then in the airport waiting hall, with no heating anywhere. My toes have never before been so cold that they have been painful to walk on. We walked out to the plane across an icy yard, which looked as though it might have been scraped with a small chisel to try to remove the ice. The windows of the plane were also covered in snow and ice, so we had no chance to see the runway. Only the roar of the propellers told us we were moving, and the bumpy earth gradually receded beneath us.

It took a total of 5 hours flying time to cover the 900 miles to Kashgar, in an old Russian Antonov two-propeller plane, armchairs with frilly seat covers bolted to the floor, luggage in a great heap in front of us. At least 45 minutes passed before the plane was warm enough for me to forget the ice-blocks that were my feet and appreciate the snowy Tian Shan mountains, whose summits we only just skimmed over. I don't think I've ever flown so close to the ground. We stopped for refuelling and 'refreshments' at Kucha – only a glass of weak tea was available at this remote desert outpost – then back on to the plane for a never-ending flight across desert sands and dust. No food was provided on this flight, just sickly hot orange cordial, two packets of sweets and a pretty pot of pink hand cream.

Eventually we started our descent into the dust, with some turbulence and no visibility; I wonder if they fly by radar here? I was very sad to miss the fantastic views of the Pamirs and the Karakoram that we would have had in clearer weather. In the airport we had the usual long wait for baggage, followed by a fairly long bus ride into town, where we were dumped in the street. No CAAC office was in evidence nor mentioned in the guidebook.

Then began one of the most frustrating afternoons. Five or six people sent us off on a half-hour walk to a hotel, which appeared to be closed when we arrived. We somehow worked out where we were on a map, found a bit of bread to sustain ourselves, and walked back for half an hour to the Renmin Hotel. The staff there just laughed at us and told us to go to another hotel; even in our exhausted state they were no help at all and we felt like screaming. We trudged back to the first hotel and at least found someone to talk to, but 'Mayo, no room here'. So we sat in the restaurant and said we would sleep on the floor because we were so tired. At last they relented and showed us some kindness; a large bowl of fried rice with meat was placed before our hungry eyes and they gave us the address of yet another hotel. Bob went off to check it out, leaving me with all the baggage.

Then a hotel man with a jeep appeared and picked up the rucksacks to take them to the other hotel. I tried to make him wait till Bob returned, but our baggage had already gone, so I was forced to go with him and hope we found Bob on the way. Sure enough, the man spotted him up the road and went running off to get him. We were delivered to the Tuen Park Hotel, famous for its smelly toilets. What a heap. The toilets (if you could call them that) were next to the pigsty, and the pigs used both. There was no running water and they charged us £6 for the room. But at last we had a room and could go for an evening stroll around the market before a most welcome sleep.

18 February
Food has ceased to be something to eat; it's just something to fantasise about in my wildest dreams. Even shish kebabs and flat bread rapidly turn into grilled fat and dry bread when there's nothing else to eat.

Kashgar is in the same time zone as Peking, even though it is thousands of miles to the west. So it isn't really light in Kashgar till 10.30am and nothing starts to happen in the streets till 12 noon. Conversely, the evenings go on till late at night, and it's great fun to wander round the market soaking up the mediaeval atmosphere. We had an incredible walk around the old streets; mud buildings, wet mud and deep puddles in all the roads, mosques all over the place, flat bread in various sizes sold everywhere, inquisitive children following us. Stares from all the adults made me feel like the prime exhibit in a zoo and it became unbearable after a while.

So we hitched a ride out of town on a donkey cart to visit the Abakh Hoja – a mini Taj Mahal in blue and green, with a strange cemetery full of mud domes and tombstones. An inspiring place, in excellent condition. We even managed to buy a good quality set of postcards of it. Another donkey cart with three ladies on it invited us to join them for the ride back to town; we slipped off the back on the outskirts in order to savour the local atmosphere with a cup of tea and dumpling in a teahouse. It seemed like centuries ago, sitting in this mud hut sipping tea and trying to converse with the Uighur men who looked at us with friendly astonishment.

Back in the hotel, we were resting and having lunch when the hotel staff barged into the room and told us to move. I was getting really fed up with their total lack of respect for privacy – the same all over China – so we finished our Wu Can Rou rolls at our own pace and went shopping for decorative local hats. Then we returned for coffee, and before going to bed found we were locked in the building.

Kashgar street scene 1985

Bob had to leap out of a first-floor window and climb down the doorframe to go and find a key to let us out. A few Chinese walked past the window but were totally unconcerned at our cries for help. Anyway, he soon returned, alone, without the key since they had lost it. He climbed back in with the help of a chair from the restaurant, and we packed our rucksacks by torchlight before both escaping via the window complete with baggage. We were then given a room in another part of the hotel, which still had power and open doors. My affection for this dump of a hotel was not improved.

Uighur man in the bazaar

19 February
The day started with the customary trip to the loo, but this time was different. With the Spring Festival imminent, several pigs were meeting a most unfortunate and cruel end right outside the ladies' toilet. I tried not to look but could not avoid noticing the poor pigs being chased with a pitchfork and then being forcibly held above a pit in the ground, screaming pitifully while their life drained away.

Having bought our tickets for the three-day bus ride to Turfan, we tried to find the Old Mosque. For some reason five different local people told us to go the wrong way. We found it eventually by ignoring their instructions and following our instincts, but it was in a terribly dilapidated state. On the way I found out what the women use instead of loo paper ... I entered a small two-hole toilet with a 2ft-high partition between the holes, and proceeded to do what I had to while the woman stared. When she'd had enough of me she blew her nose in the front of her yellow silk skirt and then wiped her bottom with the back of it. Ugh!

Deluxe accommodation in Kashgar

More wandering round the market streets with no other Europeans in sight – it's difficult to have too much of this fascinating, albeit grubby, way of life. Then we went to the Renmin Hotel to meet Jonathan after his four-day bus ride from Urumchi. He had managed to get a room at the Renmin eventually; when he first arrived he got a cold reception and was sent to another hotel, but not the one we had first been sent to. This was closed but nobody told him, despite his meeting several people on the way there. He then tried another hotel without success (also closed) and returned to the Renmin exhausted after two hours' walking. Here a different girl was on reception and she gave him a room without question. We arrived and heard the story over tea.

I suggested to Jonathan that since he had been lucky enough to get into this cheap hotel we should go out straight away before they changed their minds, but he was tired and needed a rest. Unfortunately we waited too long, for the Public Security man soon poked his head in and said he would have to leave. Jonathan pleaded that he was exhausted and the hotel man appeared to offer to carry his rucksack to the other hotel. However, after about half an hour of fruitless arguments he had to move, and the hotel man was nowhere to be seen. Luckily we had a spare bed in our room so we took Jonathan there to avoid further hassles and then went back to our favourite kebab stall for dinner.

On another evening stroll around the market we spied a grim sight. Fat white buttocks hung next to brightly-embroidered hats and painted wooden saddles; piles of sorry-looking sheep's heads adorned the muddy streets, amidst some more unmentionable parts of their anatomy. Before I came to China I was certainly more fussy and fastidious about meat than I could afford to be here if I wanted to stay alive.

Leather boot/shoe shops abounded – tall Wellington-type boots for the men and smart high-heeled shoes for the women to wear with their aforementioned silk skirts. All for walking down streets of mud with puddles so large that the streets are reflected in them, and you need stepping stones to cross. Some very lurid carpets were also on sale, but I didn't see any that I would care for in my house. Armed with three bottles of tasty Chinese beer as a special treat, we retreated to the privacy of our room for a celebration. Although with some regret, tomorrow we would be starting our journey homewards, to food, sleep, cleanliness and good health.

From Kashgar we took the bus to Aksu. The hotel had no toilets; we were told to find a place out in the yard. It was ten below zero at night and one could easily skate on all that human waste.

Our visas had almost expired, but the people in Turfan said it was too early to extend them. Two days later at Jiayuguan where the Great Wall ends, so did our visas. They did not want to let us off the train. 'Mayo, mayo, mayo – no, no, no.' It was always the same.

Jiayuguan Fort

At least there was a chance to explore this wild outpost in the remote desert region. The visas were of course sorted out eventually and we boarded the train again for Xian to see the famous terracotta warriors. It was a staggering sight and so was the mayhem at the station the next day, where people climbed in through the windows in a mad dash for space. It was impossible to sit down or even fall down! Eventually, after another sleepless night at Wuhan station, we reached a bicycle-filled Beijing. In 1985 many people still wore the traditional blue or green Mao 'uniforms'.

Bicycle-filled main street of Beijing 1985

Forbidden City, Beijing

Trans-Siberian Railway: March 1985

Life revolved around trains for three weeks, as we journeyed to Beijing and then took the Trans-Siberian through Mongolia to Moscow. We nearly missed the train from Beijing, after becoming fogged in the head with endless days of bureaucracy procuring the necessary visas and train tickets.

Ulan Bator view from the station

On the train we met two elderly gentlemen who just didn't look like aging hippy budget travellers. They seemed to be guarding their dark attaché cases with more than a hint of casual care. In the dining car we eventually ascertained that they were taking Her Majesty's Government's monthly papers to the British Embassy in Ulan Bator. Some people will do anything for a free ride!

In the dead of night the train stopped at the Mongolian border, where the wheels were changed underneath every carriage because of a different track gauge. This seemed to take hours and it was bitterly cold outside. Bob somehow got lost in the station trying to change money, but eventually we were reunited. By dawn the train had rolled and clonked its way almost to the Mongolian capital. The rolling grassy plains dusted with snow gave way to suburbs of yurts. The rounded nomadic tents seem to outnumber the communist concrete houses. At Ulan Bator's main station we were not permitted to leave the station area nor take photos of the station, although we were permitted a few photos looking down the street before the officials waved us back inside.

The journey continued on through Irkutsk, around Lake Baikal in the misty light of mid-winter. The train was warm and cosy and the scenery sped by, a mere haze of trees, snow and the occasional smoke stack. Irkutsk, Krasnoyarsk, Novosibirsk, Omsk and Sverdlovsk (now Yekaterinburg) slid by in a lethargic freezing fog. Every day we ate little else but borscht soup and burnt fried eggs. The menu's vast list had plainly been a figment of a vivid imagination or the ingredients had been sold off by the kitchen staff. The friendly waitress proudly showed us her cross, hidden in an ample bosom, proclaiming her secret Christianity.

Russia was still under communist rule then, so it came as a surprise to find that after all the hassle over our Russian visas, we were now all left to sleep at the train station when we arrived. The cleaners' cupboard provided a quiet retreat for some of us, but the cleaning ladies arrived very early and were none too pleased with us, the foreign devils from Outer Mongolia. 'Nyet, Nyet.' At least it was a change from 'Mayo, mayo'.

St Basil's, Red Square, Moscow 1985

Jaroslav en route to Novosibirsk

The next day found us trundling through Moscow, trying to change money, travelling on the underground with my schoolgirl Russian 'O' Level proving useful in distinguishing between entrance and exit, and eventually reaching the station where we could catch a train to Leningrad (now Saint Petersburg) and Helsinki. As dawn broke, we noticed the main difference in the countryside was that there were now tall buildings and supermarkets among the fir trees and snow.

'You cannot sleep here; please you come with us.' We got into the white van and trusted our luck to God. But this was God's messenger driving us. He whisked us off to the Swedish Church Mission house in Malmo. It was Friday night and the train from Stockholm had arrived too late for a connecting ferry to Denmark. The unfortunate guests in this house were mostly drunks, drug addicts and homeless. We spent the night talking to them; they all spoke English.

We're back in Europe, but the problems are the same the world over. It's just much colder here. Sleeping in a cardboard box is not an option to be contemplated as much as it might be in Colombia or Calcutta.

After yet more trains and ferries we arrived back at freezing Felixstowe. Our marathon journey was at an end and a summer of work in the Alps looked a possibility.

Put on your trekking boots

It is impossible for any thinking man to look down from a hill on to a crowded plain and not ponder over the relative importance of things.
The Mountain Top, Frank S Smythe

The Alps: Summer 1985

Summer in the Alps proved to be full of surprises, with lots of ups and downs and many different trekking groups. We were now working for Sherpa Expeditions; it was the first time Siân had ever done a job like that. It started well in Greece. Both of us were very new to this area. Ann Sainsbury was on hand to train and help us. I had first met her in Kathmandu, when we both climbed out of an Encounter Overland truck, having been collected from the Kathmandu Guest House by some Encounter drivers and taken to their house in Maharajganj, not far from the old Exodus house. It was the Christmas Day party of 1979.

Northwestern Greece has some fabulous scenery, yet it was hardly on the trekkers' map in those days. Wild dogs and rough paths, together with no proper maps, made it all a demanding and nerve-racking experience. The Vikos Gorge is the 'Grand Canyon' of Greece and surely ought to be on any trekker's list. The high ridges and airy panoramic views of the region adjacent to the Astraka hut are sensational. Throw in a climb around Smolikas, the second highest peak in Greece, and it's a recipe for adventure as good as any mountain region. The yoghurts, salads and Ouzo are other reasons to check it out even today, years later.

We were also to lead treks in Austria, Switzerland and France. It was an eye-opener in many ways. We had always thought such places were far too tame for the likes of us, but such fixed views were soon shattered. The contrast between Nepal and Switzerland could hardly be greater, except for the common theme of the mountains and the cheese. Unfortunately the season came to a rather low end, with one group being dissatisfied with their trip. We were quickly learning that not all groups react in the same way. Our inexperience in these mountains showed up when deep snow blighted a mid-summer trip, the track being completely obliterated under white powder and invisible snow-bridges. We were unwilling to lead the group over high passes in such dangerous conditions, but this was not something that certain members of the group would tolerate.

Of course today, where health and safety has gone from one extreme to the other, our actions would have brought support from the office, but back then the macho attitude was that we should have kept on going regardless. At the end of a long season, we were again out of work. We didn't actually get fired, but it was close.

Back to Kathmandu when all else fails: Autumn 1985

Being once more self-unemployed, we made tracks for Nepal in hope. Some trekking work was on offer; Sherpa had a trek over the New Year and an Australian company needed a leader for a short Everest trek up as far as Thyangboche Monastery around the same time. In the meantime, in November rumours abounded in Kathmandu that it was now possible to get into Tibet from Nepal. This was a very tempting prospect. It appeared that visas could be obtained by signing on for a three-day tour to the border town; once there it seemed that the Chinese were tolerating individuals leaving the trip and heading off for Lhasa on their own. The temptation was too great to resist.

On the roof of the world at last: November 1985

Milarepa was one of the great hermit saints of Tibet. He condemned himself to living an austere life, a lonely life in meditation seeking enlightenment. He lived in solitary isolation hidden in a cave deep in the Himalaya. One of his caves is said to exist just to the north of the village of Nyalam and below the great Himalayan pass of the La Lung La. Both places are located beyond the main chain of the Himalaya, northeast of Kathmandu and west of the great peaks of Xixapangma and Langtang.

Acrid smoke filled the room. The noodles were piping hot; outside a blast of icy cold air rattled the loose windowpanes of this shack of a café in Nyalam. Our driver beckoned us over as he inhaled the last dregs of his cigarette, the packet tossed carelessly down on to the floor. Still he was a cheery fellow, unlike some of the Chinese we had encountered before. He had agreed to take us on from Zhangmu, the border town, in his empty bus. Nobody had objected to our leaving the three-day tour on the last day instead of returning to Nepal. The bus lumbered slowly up the narrow, frighteningly narrow road from Zhangmu, through a tremendous chasm, a gash in the Himalayan mountain chain. The cliffs soared for thousands of feet above; a raging torrent as white as ice crashed about below us.

Gradually the canyon opened perceptibly. Suddenly, behind us a peak hovered into view; this was the north face of Phurbi Chyachu, a peak visible from the roof of the Kathmandu Guest House roof and most of the valley. Such excitement was tempered only by the unreal nature of this adventure.

From the noodle café in Nyalam, it was a short distance to the shrine that marked the place believed to be the cave of Milarepa. From here on it was a never-ending climb, the road switching from side to side. It grew so cold that we both climbed into our sleeping bags; it was just after lunchtime. Petrol fumes engulfed the bus; the driver lit another cigarette, but who could blame him in these conditions? He did this run for a living, a poorly paid arduous living. We felt pangs of guilt; our only consolation was that our payment for the ride was helping in some small way. The hours passed.

The La Lung La pass has two summits, both decorated by prayer flags and cairns of boulders. The wind threatened to tear the flags away. A more spectacular panorama is hard to find, except in the rest of the Himalaya. A great sense of expectation enveloped us; each new twist of the road brought some new landscape or viewpoint. Mud-walled villages, the houses decorated by fluttering prayer flags, wood smoke billowing here and there, warmly dressed children playing, a glimpse of a distant white summit, the rushing river, the colours, all shades of yellow, red and brown, the blue sky, a vulture circling above, the drone of the engine – so many overwhelming images impacted on our minds.

Just before sunset we had a view of the northern ramparts of Everest from old Tingri, across a vast plain. The north face was clearly visible. The night was spent at the new Tingri/Shegar checkpost in a very dark and dingy rest house. 'Hello, Bob Gibbons isn't it? I don't believe it, what the heck, what are you doing here, my God…' drawled a voice, an American accent for sure. It was a client who had given me a hard time on the disastrous Exodus trip I'd done with Melanie. That was the time when a truck had crashed in Kenya and we ended up in Juba, Sudan. We reminisced the entire evening, at least until our hands and feet couldn't stand the biting cold any longer. It wasn't so much better buried in sleeping bags. The whole of the next day was spent crossing high passes and bouncing along river canyons to Xigatse. The driver dropped us off at a Tibetan guesthouse close to Tashi Lhunpo Monastery. What a magnificent place.

Within the main chapel complex at Tashi Lhunpo is a three-storey-high golden Buddha, Maitreya, the Buddha to come. One can only hope he comes soon to relieve the stresses of modern Tibet. A few monks were solemnly gathered in prayer; a large drum boomed out, mumbled chants in a deep low sound droned out, the smell of sweet burning juniper incense pervaded the gloomy chapel. It was easy to be drawn into this mystical void, a chamber where one could almost feel the presence of gods, demons and spirits.

Tashi Lhunpo 1985
The monastery survived most of the excesses of the Cultural Revolution, probably because the Panchen Lama was thought to be more pro-Chinese than the Dalai Lama. New evidence now portrays this as not necessarily true. The Panchen Lama also suffered at the hands of the Chinese masters. A struggle over the succession to the Panchen Lama is still ongoing. The Tibetan-chosen incarnate, a young boy named Tukten Choeki Nima, has not been seen or heard of since being taken into custody 'for care and safety' by the Chinese. Meanwhile they have declared another boy to be the new incarnate; he does not reside in Xigatse but in Beijing.

Xigatse fort 1985

Gyangtse is a great place, the most atmospheric of Tibetan towns. It remains basically unscathed. The Kumbum Monastery did suffer some destruction but the magnificent stupa of Palkor Choide remains intact. Typical whitewashed houses, with black-painted window surrounds, line the streets and the fort has survived.

Siân below Gyangtse Fort 1985

Overlooking Gyangtse 1985

> **Gyangtse Fort**
> This fort was the scene of a siege in 1903–4, when a British expeditionary force advanced on Lhasa from Darjeeling in India to enforce trade agreements. Although the force did reach Lhasa and protracted negotiations did come to some agreement, little further trade or contact followed. Tibet at that time was living through a period of independent government free from Chinese power or other outside influences; the 13th Dalai Lama held power, while the Chinese were busy in the northeast with the Manchurians. Tibet patched up its quarrel with the British, and permitted a number of expeditions to attempt Mount Everest from its northern flanks adjacent to the Rongbuk Monastery and the tremendous glaciers that drain those northern faces. So far as we know, none managed to succeed, but the 1924 expedition, which included Mallory and Irvine, might have reached the summit. That story is still unfolding, with the discovery of Mallory's body high above the North Col, but the truth may never be known.

We decided to stay on in Gyangtse, and bade our charming driver goodbye. The final leg to Lhasa proved an inspiring ride. The route followed in the tracks of Younghusband's expeditionary force over the Karo La pass below snow peaks. It was here that a battle had raged, with British soldiers, Indian sepoys and brave Gurkhas scaling great cliffs to remove the equally brave Tibetan ragtag army in order to open the route to Lhasa. From here the gravel road cuts through a narrow defile to the deep blue Lake Yamdrok. It follows the northern shore of the lake, climbs the Kamba La pass and then drops into the long valley in which Lhasa is located.

In the time we had been away from Lhasa, some new construction had taken place. A glossy new plaza, Chinese-style, west of the Jokhang had engulfed some of the poorer areas that had previously been rather squalid, but in the process some character had been lost. The Snowlands Hotel was busy, the thermos flasks were now scratched and the bedding not so fresh. Getting back to Kathmandu proved difficult; eventually all of the backpackers heading to Nepal hired a bus. It proved a long and tiring journey. Some of the party had little patience with the driver. We decided to get off the bus in Nyalam and walk the last 25km down to the border; the scenery was well worth the trek.

Back in Kathmandu our trekking groups were due soon. But first there was a small personal matter to sort out.

A couple for the road

District Commissioner's Office, Kathmandu: 16 January 1986

Heavy chains clanked on the floor. A pervasive aroma of urine practically killed all sense of smell. The two men were chained together and looked pretty down-at-heel. The District Commissioner had evidently given them something to think about. They were ushered out of the room we were about to enter, destined for some dreadful airless prison cell no doubt. As for us, we were here for the legal formalities. The D.C.'s office performs many different tasks. Our thumbprints were added to the papers; the ink stuck fast and would not wash off for a couple of days. Indian Airlines flew noisily overheard as we made our vows in Nepali, we hope! In less than ten minutes we left the room, pronounced husband and wife with an elaborate rice paper marriage certificate. We were not leaving chained together.

The DC's Office, Kathmandu – the newly married couple

In fact this short ceremony had already been delayed by a week when Siân was offered her first ever Everest trek. She flew up to Lukla, the dramatic airstrip for the short Everest trek, two days before the original date we had planned for our wedding, spending the actual day resting, acclimatising and contemplating in Namche Bazaar. A small reception was held in the Astha Mangal Tibetan restaurant with a few friends; the total cost was £7. We later had a party in England for all the family.

Siân off to the reception at the Astha Mangal

By way of a honeymoon we had a night in the Everest Sheraton hotel courtesy of Chris and Tashi Lachman from Australian Himalayan Expeditions, whom Siân had just worked for. A couple of nights in the Khetri House hotel in Jaipur followed, on the way back to Delhi and England.

Married life proved very agreeable. In the spring Australian Himalayan Expeditions (AHE) offered us both trekking work in Nepal, followed by an exciting reconnaissance trip in Tibet. The plan was to find a trekking route from Tingri to Rongbuk and the north Everest Base Camp. This proved to be a relatively easy trail, except for the problems of high altitude. We met a young French couple coming down, who offered us homeopathic coca tablets to help against altitude sickness; we tried them and have used them ever since.

Base Camp was full of expeditions preparing for spring summit bids. Tibet was firmly on the climbing and tourist scene now. Over the next year we worked in the Alps and Nepal for AHE. We did some new routes in Nepal and one trek in Tibet. Over Christmas and New Year we were sent to the Far East to collect masses of trekking equipment for the amazing increase in business that the company was experiencing that year.

A window on Yemen: March 1986

One country we had long wanted to see was the Yemen. In 1986 it was still divided, but the superb mud architecture, the mountains and markets were a revelation. The capital Sanaa was still a very traditional Yemeni-looking town with gaily painted, high-rise tower houses. The people were incredible, the men swaggering about with huge curved knives and guns that looked as though they would never work. The southern circuit by local bus took us to Ibb, Djibla, Taiz, Hodeidah and back to Sanaa. Out to the east is the ancient dam of Marib and its crumbling citadel.

Sanaa (southern gates), Yemen 1986

Characteristic buildings in Sanaa

Sanaa market

We took a bus to Saada in the north, a fabulous place of intricately designed mud-walled tower houses, wild-eyed country folk and kids not so politely trained.

Saada – the northern town in Yemen

Our resultant small stapled guidebook to the Yemen did well until Lonely Planet finally issued a guide to that country.

Wadi Dhar Palace, Yemen 1986

The tragedy of Yemen
Yemen is a much-underrated country. It developed a significant tourist industry until kidnapping and finally, in 1998, the murder of some tourists put paid to all further progress. Sadly we never got to the Hadramawt with its amazing sights at Shibham, Tarim and Seyun. And probably never will, if the constant bad news today is anything to go by. The Arab Spring affected Yemen when its long-term dictator Abdullah Ali Saleh was forced into exile, but it did not produce the hoped-for results. Houthis from the Shia north invaded much of the country and then the Saudi-backed bombings reversed some of the gains. Throw in the Al Qaeda and ISIS groups, and the place is a melting pot of terror. Whatever the truth of the fighting, it is the ordinary people, as always, who are the losers.

Chamonix: a pied-a-terre in 1986

Following our wedding in Kathmandu, we knew that we would never be happy living in London, so it was time to sell my small two-up/two-down house, which I had rented to the same tenants since moving out to the east with Bob in 1984. We were in the fortunate situation of being able to live with Bob's parents while we were in England, and we had both fallen in love with Chamonix, so that seemed the obvious place. We found a newly built one-bedroom apartment in a chalet facing the Dru, an iconic peak in the Mont Blanc range, and that was it. We would make it work somehow! In the summer we would use it ourselves while working in the Alps, and in the winter we would rent it out to ski reps while we worked in Nepal.

The Dru from our balcony, Chamonix

Unemployed and re-employed again

Our Himalayan trekking bubble burst in the spring of 1988. The Australian company suddenly appeared to be overstretched; we were all laid off and a finance company picked up the pieces. It later became World Expeditions. Luckily in the autumn some trekking work in Nepal came our way from Explore, a British company. Siân's parents came to Nepal for their first trek, around the Kathmandu valley rim, while Bob was leading elsewhere.

The volume of business in Alpine trekking had been improving, so we returned to work for Sherpa Expeditions once more. From 1989 onwards we were to spend every summer in the Alps, mainly in Switzerland, operating the Alpine Pass Route in the north from Gstaad to Engelberg and the Haute Route from Chamonix to Zermatt. It was a great job; it needed at least two staff and often more. The summer job became our main source of work and funding for our travel. Sometimes there were other trips to Tuscany, Provence, the Loire, the Vermillion Coast of France, the Spanish Pyrenees and Greece for research. The job started to seem like a 'proper job' as it continued for more than 30 years, every summer.

There was always plenty of work in Nepal too, until 1992 when things took a downward slide. Increasingly local leaders were being employed, as was only fair. The trekking scene had blossomed since my Exodus days into a very good business. Nepal was now earning more from tourism than any other activity.

Many other things were changing in Nepal as well. The traffic had been steadily increasing; the population had grown from 8 million in 1975 to close to 23 million by the new millennium. The standard of living had not been growing, though. The old Panchayat system of government had survived a referendum in the eighties, but it didn't look as though it would last much longer. The Panchayat system operated through a collection of small councils from village level and regional to a centrally elected body. In principle, the king was a near-absolute ruler with ministers and civil servants concerned with day-to-day affairs. Things were unravelling fast.

Facets of West Africa: 1990

In 1990 our enthusiasm for wild places took us to Togo, Ghana, Mali and Burkina Faso. The heat of West Africa, even in its winter, is certainly a challenge. Imagine living in that climate all year round! The early highlights of the trip were the fetish markets of Lome, the sombre slave forts along the Ghanaian coast and the vast, overwhelming market of Kumasi. North of Ghana, the secretive nature of Burkina Faso, after its emergence from the brutal antics of its former dictator, was a little unnerving. The country people were very friendly and the country later became a delight for adventurous travel. In Mali the astonishing spectacle of the Bandiagara escarpment was a hot paradise for trekkers. Along the base of the cliffs are the famous Dogon villages.

High above in the cliffs are some curious granaries, sacred spots and amazing rock art. The cities on the banks of the mighty Niger River hold untold treasures. Here are the amazing mud mosques and secretive alleys of intriguing Djenne. Nearby is the equally mysterious settlement of the Bozo. The famous mud mosque of Djenne is held up by hundreds of timber beams and every year new mud is added to the outside after the rains. It's a very evocative building; a picture postcard image of West Africa. Upstream is the bigger town of Mopti, fishermen and market traders crowd the shores of the river, where a shabby-looking steamer is about to depart downstream at any time but the present.

Segou, once home to the French colonial administration, is another sleepy place where little disturbs the daily rhythms. Dust blows up in the main street, children scurry around strangers in town, while the call to prayer emanates noisily from the minarets, crackling and unintelligible to outsiders. The brash and breezy capital Bamako is not much more than an overgrown village at heart, with miles of thatched hut suburbs surrounding the hardcore centre of concrete. The central market sadly burnt down a couple years after our first visit, never to be reincarnated. There are easily enough surprising spectacles to add spice to the mix of a journey in West Africa.

Dogon villages below the Bandiagara escarpment

Dogon dancers, Mali 1990

An enthusiastic welcome in Ghana

Ganvié stilted village, Benin

African sunset 1990

On our return more work was on offer in Nepal for the spring season, including a possible recce of the Manaslu region. Siân went out earlier as Bob had some family matters to sort out in Chichester.

Kathmandu Under Curfew: April 1990

Bob's report

Siân has been gone now 10 days now on trek in the Everest region, but here I am at Heathrow and all is on time. Eighteen hours later, I touch down in Kathmandu – it's stormy and raining and trouble is brewing fast. At the airport Pablo is waiting for the Kanchenjunga group – there were demonstrations on Monday, with the army opening fire on groups from helicopters. Some government ministers have been sacked – the opposition leaders were released after being locked up, some say after being made to stand in water. It doesn't sound like Kathmandu at all. On Thursday things are looking quite bad; there is to be the biggest demonstration of all tomorrow and we are being told to be off the streets tonight by 6pm. Just time to see Ravi and have dinner at the Moti Mahal with Dilip Singh. It seems I was on the last flight in from London on Biman Bangladesh Airlines. PIA came in later that afternoon. A day later Biman was to get just one more flight out, on the Friday morning, the day it all began.

Friday 6 April
I leave the Kathmandu Guest House to get breakfast in the Mona Lisa café, but all of Thamel is closed and no one is serving food. People are loitering in the streets and army units are down Tridevi Marg by the supermarket bazaar. I return to the hotel; the Astha Mangal restaurant is closed. No breakfast. Everyone in the hotel is in a state of nervous anticipation. Eventually I decide to try to get to the Yellow Pagoda Hotel, nervously walk down to the roundabout and along to the hotel, deciding to get lunch in Kebab Corner. I talk to a South Indian fellow working here.

At the hotel Andrew Bluefield is there; Pablo and Holly are out with the Annapurna group at the airport (travelling by cycle rickshaw, as there is no other transport). Holly phones from the airport and requests me to try to contact the Biman office, as there is no one to check the group in at the airport. Andy comes along to check Nepal Express. Things don't look too good down Thamel way, so we hurry on past the Royal Palace to Durbar Marg. The Biman office is closed, although we saw the fellow go in and pull the steel shutters down. We look back to Thamel. A mob is gathering, with army units facing them off, so we head back down along Durbar Marg. People are gathering near Rani Pokhari lake; a large contingent of the army is nervously waiting on a truck. We get back to the Yellow Pagoda.

The demonstration begins. Holly phones again to say the Biman aircrew loaded the luggage and the flight departed quickly. I warn them to be careful getting back to the hotel. From the roof of the hotel we are watching huge crowds of demonstrators; over an hour of continuous, chanting people pass by. A tourist is seen marching, others take photos below. The army unit by the Hotel Nook has retreated down the street by Kebab Corner. Pablo and Holly arrive back via backstreets, passing burning tyres. By 3.30pm loud bangs and gunfire can be heard in the direction of the clock tower and New Road.

I decide to get back to Thamel quickly. It's very nerve-racking with an eerie silence, groups of army and locals standing off. I hastily pass into Thamel and back to the hotel. No one knows if it's safe in Thamel, but we try to find food. I follow a German tourist down the street; the army has met with demonstrators who are running my way. I turn and run back for cover, back to the Kathmandu Guest House. It was a teargas attack by the army; rumours abound of shots and casualties. I buy water from the KGH shop. At reception people have carried baggage from the airport through teargas attacks to reach the hotel. They are quite old and very shaken. Stories abound.

On the roof gunfire can be heard frequently, fires and smoke can be seen in Durbar Square and New Road areas – cars or tyres. There are fires in Dilli Bazaar, Teku and Lazimpat. None seem particularly big. Burning rubber smells fill the air; it's twilight, the sky is red over Swayambhunath. A tourist has a radio and is listening to the BBC World Service. They are saying people have been killed – ten here and fifteen in Patan. As we listen, gunfire rings out; the sun is redder. An ex-minister is interviewed on the radio. More new fires, we are all hungry. Someone says they are serving food in the gallery room of the hotel. We check it out. The Tibetans from the Astha Mangal have set up a kitchen in the Gallery Room.

Back on the roof we can see a group chanting below with lit torches – they are marching towards Chetrapati. Some new fires are visible less than 100m away, more burn across the city. More gunfire is heard. Down in the gallery room, Hari from Natraj Tours is organising a second sitting for dinner in a robust manner. We get rice, vegetables and cold drinks – coke and beer. Soon the rice is finished, but the noodles are going strong. It's an electrifying atmosphere. People are trying to call home. Local calls are much harder. Everyone is talking, relating their version of events. It's quite a spectacle. It's 9.30pm and I return to the room, we have to keep the lights out. I try to boil some water for coffee. It's hard to sleep; my brain is too active.

I wonder where Siân is now – maybe Phortse? Does she know what's happening? M.A. (Harper, another leader) and her group are in Lukla without tickets. Holly's group should be in Everest or rafting by now, but are stuck in the Yellow Pagoda. Andy Bluefield's Kanchenjunga group flight is off. What does Siân know of this? The waiting game begins; will any flights operate tomorrow?

Three times I go to the safe and collect money, passport and medicines. I pack to go trekking, in case I have to run for the mountains. I plan not to leave Kathmandu until Siân is down, or I have to go to Jiri to meet them if they walk out. How can anyone make a plan for this situation? It's still hard to sleep, but there is no interruption after 11pm. Some press people from the UK apparently woke people up to check stories about British casualties – some were killed or injured on Durbar Marg around 4pm – just 2 hours after we were there. It's very disturbing. Should I try to get out to Lukla?

Saturday 7 April
I wake early, but don't get up until 8am. I don't really want to know what's happened. The Astha Mangal is open but nowhere else. I walk out towards KC's, but the army tell us to return to the hotel – there's a curfew. Barriers have been erected by Coppers Café. Army units are everywhere. I get some scrambled egg and tea in the restaurant. It's crowded with tourists. Others are out in Thamel looking for water and food. People gather in the KGH reception. More stories abound. Curfew is now imposed from 7am to 4pm today. Someone has a big radio in reception. I try to phone the Yellow Pagoda Hotel, but the lines are poor. The Voice of America refers to the problems here: forty killed in Kathmandu, fifteen in Patan. It's very vague. Some German tourists are very agitated about their flight back on a Condor charter. It seems their plane is waiting in Sharjah. Others talk of walking to the airport and camping there. It sounds crazy.

People ask about flights and news. Raju from Encounter is tuned into the army waveband. Now on the BBC World Service is the 5am GMT news – more casualties are reported during curfew this morning. Should I phone Derby yet? There is a waiting time of one hour to call out from the Kathmandu Guest House. Finally I get through to Holly at the Yellow Pagoda and Jim, from Sherpa, is phoning regularly from London. The British Consular Official Mr Mason has been at the KGH this morning, checking out the details of the British tourist who was shot. There is another in the hospital with serious leg wounds (both were shot near Durbar Marg). Many people are asking many questions, mainly 'will they evacuate?' Mark Tully of the BBC is heard on the radio in the courtyard.

It's time to queue for dinner; no one has eaten since breakfast. It's a long day in the room or on the roof. It's so quiet. Holly and Pablo are stuck in the Yellow Pagoda. Downstairs in the Astha Mangal restaurant, a girl has fainted in the doorway. We eventually get rice, vegetables and coke. Once again Hari from Natraj is officiating. I talk to a New Zealand fellow, delayed leaving with Encounter, and a couple hoping to trek in the Pokhara area. There seems to be talk of some emergency evacuation flights tomorrow organised by the Nepal Hotel Association – either to Bangkok or Delhi. Many people are signing up in reception. Upstairs the phone is going all day, now eighty-five calls made by 9pm. Once more on the roof, it looks quiet. Time to retire but hard to sleep.

I think about Siân, what do they know in Lukla? She must be in Thyangboche or Namche now.

Sunday 8 April
What will it be today? I'm up early and it seems okay to go out. Quickly off to the Yellow Pagoda; I think it's okay. There is a rumour that the curfew is lifted but then it suddenly seems only until 8am. It's already 7.45am at the YP hotel and they have a room. I decide to stay there to keep in touch with Holly, Sherpa and Himalayan Explorers. I borrow Pablo's bike to race back to the KGH to collect my luggage by way of Tridevi Marg. It's too risky to return that way past the Palace, so I take the backstreet way through Jyatha Tole. It's dodgy cycling with one hand on the bike and the other holding baggage. Back at the Yellow Pagoda, it's time to cool off. Coffee and breakfast with Andrew Bluefield.

Curfew seems to be lifted now until 9am to allow government workers to get to work. We head off around to the airlines, passing shattered windows and bent ironwork by the KLM office. President Travel is a mess; people are standing outside the Thai Airlines office. The guards don't seem to know if we should be out or not, so we retreat to the hotel again. The street outside is quiet now. I spend over an hour talking to a couple who used to work here in the 1960s, so it's a fascinating discussion. He is gloomy about the prospects. No one at the Yellow Pagoda seems to know about the emergency flights. Many tourists are heading out to the airport on foot, against the curfew. The soldiers are not doing anything to stop them. The airport must be full – there are stories of people camped out there with little food or water.

I'm now outside the hotel in the street for a few minutes – a blue airport bus has gone by carrying passengers for the rumoured Delhi flight. There's an American judo instructor living in the annexe and a group of tourists from Thamel talking about the flight out. It seems okay for tourists to wander about, but no one really knows. Three Sherpa clients now decide to get out on the Delhi flight. There seems to be a lull in the curfew until 2.30pm, so I am asked to take them up to the British Embassy, which has been around telling people to get to the embassy quickly if they want to leave and the embassy will transport them to the airport. We set off for the embassy; a rickshaw has taken one with the luggage.

I am talking to a South African, who's familiar with such scenes and is not happy. People are getting very agitated, especially near the Palace. Armed guards with enormous machine guns by the truck-load are stationed everywhere by the Palace, and inside the compound of the Kaiser Library. I am getting very nervous about this situation so close to the Palace. Now by the Ambassador Hotel there are very few people out, just a few younger tourists, crazy.

I don't go all the way to the British Embassy but point the clients to it, just a few yards up the road, and hurriedly return down past the Palace, hoping no locals approach the scene. It's a relief to get back to the hotel. It's getting close to curfew again. Later a British Embassy Land Rover pulls into the hotel to collect others for Delhi.

It's now 3.30pm and the streets are deserted once more. Up on the roof of the Yellow Pagoda, people are watching for the evacuation flights by the Nepal airline, R.N.A.C. Two have gone so far, one plane making a shuttle service roundtrip. No one comes in. A small Twin Otter flight has come from Pokhara with some French people. There are stories of four deaths in Pokhara and big trouble rumoured for 4pm today. Andy is worrying now about his clients, about the Terai towns en route to Kanchenjunga by road. His agent has returned the clients' international air tickets in case they have to leave. Snow reports are bad; it's down as far as Lukla and settled in Khumjung. It's raining here now. Another flight is heard around 5pm. Curfew is again lifted between 5pm and 6pm. I head over to Kebab Corner for an early dinner. It's quick – Malai Kofta and naan. Our usual waiter is still on duty, he's been stuck in the restaurant now for three days, unable to get home. He used to work in the Blue Star Hotel many years ago.

Back to the hotel by 6pm. The rain is heavy. Now I meet James Roberts, who is working full time with Jim in the Sherpa office for a while, but also leading at times. He has been in Morocco; now he is out in Nepal learning the ropes here. He's staying with Graham Bullock, also working for Sherpa at present. He's been rafting with just one client – an American girl – and has just got in from Chitwan, after being stuck out at Thankot by police for a while. One group of tourists are said to have been stuck at Naubise for three days without a hotel. We are just sitting talking in the lobby. Another flight is heard – that makes four.

No one has come back from the airport yet. I keep thinking that Siân and M.A. might have to walk out from Lukla, a 10–15 day hike to Kathmandu. I talk to Jones and Stitter, a comical pair who have been trying to fly to Lukla for a week. There is nothing to do but drink. The two lads have a letter for Siân and keep trying to get me to take them to Base Camp. James is keen and may go. Then there is the Manaslu recce idea to think of. We all go off to Kebab Corner: Holly, Pablo, James, the girl and the lads. The same waiters are very helpful, but the lads are confusing them over their orders. Back late across the dark empty streets to the hotel. Andy is with a couple of clients and is looking anxious.

Finally at 11.15pm I go to my room. It's getting really grim; not far away I can hear whistling and firecracker noises, then two gunshots ring out. Now I'm really getting scared, as tomorrow is to be 'Burn the Constitution' day. It seems inevitable that there will be a lot of shooting, as people are going to defy the curfews. It's getting very noisy out there, with gunfire and crackers. I creep low around the room to turn off the light as the whistling is very near.

About midnight the people downstairs in reception say that the King has announced multi-party political reforms – it sounds crazy at this time. It seems that firecrackers are going off to celebrate; people have been shot, the army doesn't know what is happening. Music can now be heard. Is it really true? They say Mark Tully is broadcasting from the Kathmandu Guest House on the BBC World Service, it's a rumour at present. I sleep better.

Monday 9 April
It's true; already at 7.15am the streets are busy. It's incredible, everyone is excited and relieved and bursting out with joy; locals, staff, tourists, everyone has a sense of incredible relief, it's sunny, it's so incredible after the days before! No one can believe it. Already people are celebrating in the streets with flags, vehicles, cars, cows, chanting, singing; it is incredible. I get around to the KGH to book back into room 51. Raju and all the staff seem very happy, handshaking. Yogi is there with Bindu (Mrs Sakya), and Mr Singh the fortune-teller is back. Back at the Y.P. James is having a late breakfast. I feel relieved; at least flights to Lukla should start now. I am with Holly and Pablo. Stan (Armington) comes in to relate his stories about what happened next to the Palace between the mobs and the army. Maybe Siân will fly out on time?

I check the fax and Biman; we are still wait-listed for Monday 13 and 20. M.A.'s flight is temporarily off (later she has been confirmed on 13 April with us). We may still be able to do the recce. I go to see Ravi and get a telex out to Derby via the Chichester telex people we use. What do they know? Lunch at Kebab Corner where everyone is talking politics – the man from Madras is there again. Back to the hotel, Lukla flights to resume. It's a long day of celebration in the streets.

Tuesday 10 April: Kathmandu airport
I've been here now since 9am with Andrew Bluefield and his group, who are chartered to Taplejung OK. It's clear and sunny. The first flight to Lukla has gone out with full loads. About 10.30am it's

back, but with a Japanese group. Oh dear! Then the second flight is landing... I see her feet first, descending the steps to the tarmac... it's Siân's group, on time. It's amazing! ...

Siân's report

Saturday 7 April: Namche

We arrive in Namche after our aborted Gokyo trek (far too much snow), and head for the restaurant, where we first hear of the troubles in Kathmandu. Some of Per Lindstrom's balloon group are here, with a radio, waiting for the crew to bring up their balloon, in which they hope to fly over Everest and land in Tibet. The release of the balloon has been delayed by the Customs at the airport in Kathmandu, and no doubt will take some time, especially now. We discuss the situation in as much detail as we can, but of course nobody knows much.

I wonder, where is Bob? I last spoke to him in Chichester two weeks ago and know only that he had been scheduled to fly into Kathmandu last Wednesday, which is when I am told that the airport was closed. Did he get in or not?

Camp in Phortse, Everest 1990

Over dinner, I try not to show my personal worry and concern. Two female Army officers happily inform the group that they have been trained in riot control and will be able to look after the situation in Lukla airport. It appears that the airport has not been functioning for a week now, and the backlog of tourists waiting to fly back to Kathmandu must be overwhelming. It's hard to sleep, but we still have a way to go...

Sunday 8 April: Phakding

None of us has slept much, and now we wonder what will be the first commodity to run out. What could we least do without? The first thought that springs to mind is toilet paper; with so many Westerners stuck in Lukla, surely the local shops will be out of stock? So along the way, in every tiny box-like store, we purchase toilet rolls. It's not far, but it seems like forever and I just want to get back to Lukla. Tomorrow we will find out the truth.

Monday 9 April: Lukla

We wake to hear that the king has announced multiparty democracy. The airports will reopen and flights will resume tomorrow. Can it be true? The skies are thick with turbulent cloud and foreboding as we climb the final hill to Lukla. How many people are stuck here? Will it be chaos? I reassure the group that if the system operates as usual, we will be first out, because we are scheduled on Tuesday morning's flight, and if you are on the waiting list you go to the back of the queue. But am I fooling myself? This is no ordinary situation.

At Lukla I meet M.A. in the kitchen of the Hotel Himalaya. We sob a little on each other's shoulders as she tells me what she's heard. Some foreigners have been killed in Kathmandu, but I feel sure that Bob will not have been out on the streets if there is a curfew, if indeed he has reached Nepal. I try to reassure myself that he's okay. The last night party does not have its usual ambience, but we try to relax and hope for the best. At least there's no shortage of toilet paper.

Tuesday 10 April: Kathmandu airport
Up at 6am; we have not of course slept much. Thick grey clouds blanket the sky, and in this mountain airport, a mere strip on the hillside, planes cannot fly in or out if there is a spot of cloud obstructing the slope opposite. So even if the airport is officially open, the planes may not be able to fly. But ever optimistic, we wait in hope. The first plane comes in about 9.30am, but our name is not on it. We wait… and another arrives.

Take-off from Lukla before the tarmac was added

It's ours!! The emotion is indescribable. We take off, but instead of taking the normal route southwest, we turn and head north for the slopes of Everest. The view is breathtaking, and I feel so emotional I can't think straight. The mountains are so close, I must take some photographs. It seems like we're going to crash into the South Face, but what the hell, if we're not, or even if we are, I may as well take some pictures. Anyway it will help to take my mind off the terror. But over Namche Bazaar, we circle round and head south. Adrenalin is racing through my veins and my whole body is on fire.

Arriving in Kathmandu the whole situation is unreal. I can't wait to get off the plane. Is Bob there? I don't have to wait long, as he has somehow got through to the plane parking area to meet me. As I see him, even more adrenalin rushes through me and we cling to each other in our first ever public embrace in Nepal… and I don't care who's watching.

Rheumatoid Arthritis onset: July 1991

I have added the following section simply to inspire anyone diagnosed with RA not to give up hope. With luck and the right treatment, almost anything may still be possible!

It was while waking in my tent in the campsite at Grindelwald in 1991 that I first noticed a stiffening of my fingers. It soon went away for the rest of the day and I thought no more of it. Then a few months later in Nepal, my thumb was so painful that it felt like I had broken it. That also went away and did not return. In February 1992 one night I awoke in agony at 2am, feeling that I had fractured my wrist. It was below freezing and we were sleeping in our Land Rover outside the closed campsite in Joigny on our way back from an aborted trip to Algeria. I could not explain this excruciating pain, but again it went away after a day or two, so I ignored it.

Having failed to reach Algeria and the Sahara, in March we flew to Central America to explore Tikal in Guatemala and the other Mayan cities in Mexico and Honduras (see below.) I found myself unable to climb the pyramids because of pain in my legs and knees. We visited our friend Ed in Florida; he took us sailing in the Bahamas and I couldn't climb back into the boat because of weakness in my arms. Clearly something serious was going on, but I tried not to panic.

Blood and other tests back in the UK revealed that I probably had rheumatoid arthritis. It was a horrifying moment when my GP said to me that I would have to change my lifestyle and job. Not likely, I thought, determined to carry on. While on the waiting list, the pain got worse, all over my body, and would not abate even with painkillers. I found it hard to get up out of a seat and impossible off the ground. I could hardly bear to put my feet on the floor in the morning or after sitting for a while.

Not a good prognosis for someone whose life revolved around trekking, camping and the outdoors.

But on 1 May 1992, when I finally saw the consultant Dr Martin Ridley, relief was at hand. I started on Salazopyrin, which was effective until 2004, when we were driving down through Africa. Suddenly I found it difficult to get out of a chair again. I was then prescribed Methotrexate in varying doses, along with Leflunomide after a tendon rupture while driving in northern Mali in 2008. My feet became worse but were solved very simply by larger shoes and orthotic insoles, which have kept me not only on my feet but trekking even at high altitude! Eventually, though, the tablets were not enough to relieve the inflammation and in June 2014 I had my first infusion of Rituximab, to be repeated 9–12 monthly.

Currently the combination of weekly Methotrexate and annual Rituximab infusions seems to be working well and I am relatively pain-free unless the baggage is too heavy!

Mayan Odyssey: March 1992

As mentioned above, another short trip took us to Central America to see Tikal in Guatemala and the other Mayan cities of Mexico and Honduras. Taking a cheap flight to the tourist centre of Cancun, our first days took us into Belize. Arriving after sunset did not enthral us after reading warnings in guidebooks about the general crime levels, but nothing untoward overtook us.

Heading inland from Belize City, a rough road led into Guatemala. This frontier area had suffered from a lot of lawlessness and rebellion, but in the event our trip was trouble-free. The sensational ruins of Tikal need no introduction, as the area sees thousands of travellers and tourists today. But at the time of our visit in 1992, security issues in the jungles surrounding Tikal meant a short flight was needed to get to Guatemala City.

We found a cheap room in Guatemala City, but the area was pretty dire. We needed all the cupboard drawers to block the dodgy door from the inside, barricaded in but hardly daring to fall asleep. We didn't dare use the shower either, as a bare live wire lay directly underneath the showerhead. In the chaotic bus station the next morning it required our wits to be on full alert, with hustlers eager to part travellers from their possessions by many tricks. Once away from the city limits, however, there were no such worries getting to Copan in Honduras to see more famous Mayan remains.

Tikal, Guatemala 1992

Siân at the Copan ruins, Honduras

After some wild highland towns like Santa Rosa our route took us to the steamy heat of the second city of Honduras, San Pedro Sula. Some pretty dodgy characters seemed to hover around the virtually defunct town of Puerto Cortez on the northern coast, hardly the tropical paradise we had expected.

Another short hop by plane got us back to Belize and the Mexican bus to the final jewel in the Mayan crown, the vast site at Chichen Itza. The whole area was quite deserted of visitors in those days, so the atmosphere proved tranquil, slightly mysterious and inspiring. The charming city of Merida was a relaxing revelation. A few travellers were trying to fly into Cuba from there.

Belize City 1992

Chichen Itza, Yucatan, Mexico

146

Mexican Cartels
The more recent images of much of Central Mexico, north of the capital, is a sad indictment of the struggle to control the illegal drugs trade. None of this was evident in the Yucatan, though. The horror stories do not seem to be abating, although some of the troubles stem from the migration of people from Central America where places like Honduras have seen worsening security situations. Ironically El Salvador, which Bob visited in 1977, has improved immeasurably, so there is always a positive to be found amongst the negatives it would seem. Uruguay has tried a different tack, legalising some drugs. It remains to be seen whether that approach will lead to a better outcome.

Mustang: September 1992

Two friends, Gina and Peter Corrigan, approached us with the possibility of running a trip for some of her friends to Mustang in Nepal, which was about to open.

Mustang lies to the north of Kagbeni and Jomsom on the Annapurna Circuit trek. It is a land of Tibetan culture and monasteries, cut off from the world since the Chinese invasion in Tibet. It remained off limits until 1992, probably because of tension between China and Nepal, and because the territory was thought to be a hiding ground for Khampa guerrillas seeking freedom for Tibet. There were five of us in the group, and after a scary flight into Jomsom in thick cloud, we trekked into the new territory. The route kept west of the Kali Gandaki through Chele, Geling and Tsarang to the lost city of Lo Manthang.

As with most groups during this first season in Mustang, we were privileged to meet the king and queen and take tea with them. The tea was Tibetan-style butter tea, but it was very palatable compared to the tea we'd had in Tibet. All in all it was a very exciting and enjoyable trek; our return route to Muktinath took us through the bewitching landscapes of the Narsing Khola in eastern Mustang.

Tange chortens, Upper Mustang, Nepal 1992

****

The start of 1993 saw us finally get to Algeria with our new but old Land Rover. Our route included Morocco, Western Sahara and then across Algeria via Timimoun and Ghardaia to Tunisia. This was followed by another busy summer in the Alps to boost the coffers.

We finally came unstuck in the winter of 1993 with no more trekking work that season, as Sherpa moved to employ local leaders as well. We started to look for new pastures, the Sahara perhaps. In the meantime, Cambodia was opening up a little, following the UN departure after years of internal conflict and the ravages of Pol Pot.

Mind Your Step In Cambodia: November 1993

In the eerie twilight, only the haunting sound of whining cicadas broke the otherwise deafening silence of the jungle. Then a solitary gunshot rang out. Scared witless, we immediately started the motor and rode our machine as fast as possible on the broken track through the massive gate structure. Black stony faces stared down as if to say, 'You dare to come this way.' We retreated back to the central monument – the Bayon.

We were visiting the vast temple complex of Angkor in Cambodia, the remains of the ancient Khmer Empire built in the 7th–11th centuries.

18 November
The 5am minibus to Bangkok airport did not materialise, so at 5.20am we gave up and headed for the local bus no. 59, determined to get back the 100 baht we had paid for the minibus on our return – the local bus cost 3.5 baht each! At this time of the morning, the traffic was still moving freely out of the city, so we were at the airport by 6.15am. We decided to have breakfast at the airport in case we didn't get fed on the plane, so ended up with two breakfasts.

The aircraft was a Canadian de Havilland Dash 8, modern, clean and only 20 minutes late. Eastern Thailand below was very developed, no jungle in sight. Everywhere were neatly laid-out fields with crops growing, and jetties along the coast with small ports. The Cambodian border was very clearly delineated where we crossed it at the coast and headed inland; on one side the neat fields/rice paddies of Thailand and on the other, untouched jungle. The first signs of human habitation/development were the shimmering rice paddies of Phnom Penh.

Phnom Penh, Cambodia 1993

On arrival at Phnom Penh, we walked under the tails of two large jets to reach the terminal building. People were getting one-month visas on arrival. A young man met us at immigration and took us in his taxi to the Capitol Guest House. Our room, with fan, private shower and toilet, cost US$8. The hotel was in an unsurfaced side street off the main street, Achaman Boulevard, and surrounded by shops selling shiny new bicycles. It is definitely the backpackers' centre and it would appear that a lot of expats and UN personnel visit the restaurant for cheap beer!

After unpacking we went for a walk to explore the city and buy air tickets. We seemed to walk for hours; 'mad dogs and Englishmen go out in the midday sun' would seem a fair description of us, and I think we were the mad dogs! There are a lot of shops here selling all sorts of goods, and, of course, Japanese generators for the all-too-frequent power cuts. We found supermarkets selling European food and even Vegemite! Eventually we found Kampuchea Airlines and were able to get tickets for Siem Reap.

Phnom Penh
It was staggering to see the obviously rapid growth in consumerism in Phnom Penh. Everything seemed available; Thai bicycles, Japanese generators, electronic goods, European foods and Australian vegemite left from the departure of the UN contingents just two weeks previously. The headlines in the 15 November edition of the Cambodian English language newspaper seemed to confirm observations, with headlines announcing 'Growth Corridor'. This corridor refers to the areas adjacent to the main road to Kompong Som (former Sihanoukville.) A headline on the back page read 'No conflict'. A sign of the times in Cambodia, where a very tenuous peace is balanced between growth, improving living standards and a retreat into endless sporadic fighting as the negotiations drag on between the new government and the Khmer Rouge.

19 November
Breakfast was greasy eggs and acceptable but dry toast with tea/coffee for two. Visited the Grand Palace and Silver Pagoda (which was closed) and back to the Capitol for a greasy lunch. We also bought some French-style bread to eat in our room, smothered with delectable Marmite! Who would believe dry bread and marmite could be so scrumptious!

It really is so hot and sticky here. Out for a short stroll to wake up before dinner, we got soaked in a lightning storm and retreated to another plate of greasy noodles with a fishy flavour. The streets seem to have no drainage and flood quickly with the slightest rain. We met some very young UN workers having a beer in the Capitol.

21 November
This was to be one of those days we'll never forget, quite incredible. As we entered the plane, a Tupolev jet with Russian pilots, clouds of water vapour gushed out of the air vents. The flight was fairly uneventful, crossing flat countryside followed by the vast expanse of water of the Tonle Sap, before touching down on the long hard runway at Siem Reap. We met a French cameraman on his way to film the de-mining operations in the jungle around the monuments of Angkor.

Not wishing to miss any opportunity, Mr Proen Prav, manager of the Mahogeny Guest House introduced himself outside the airport and guided us gently to a waiting taxi and into town to his guesthouse. His lovely old wooden house with elegant teak upstairs veranda had recently been the home for UN suppliers – Morrisey Supplies. He told us that a few months ago the Khmer Rouge had terrorised the neighbourhood adjacent to the building but that things now seemed OK. It was also possible to visit the ruins without guides and hire small motorcycles for the purpose.

The main temples of Angkor are about 6km north of Siem Reap. We hired a motorbike for US$5 and set off for Angkor Wat. Such excitement was almost too much!! Along the route were a number of burnt-out houses and a couple of tanks were hidden on a side track. We followed an overloaded motorcycle for a short distance on the way to Angkor Wat. A live pig was strapped unceremoniously across the back seat of the machine, its legs waving pitifully in the air.

Angkor Wat is a truly amazing place. Surrounded by a vast moat, inside is a very green open area with cows grazing, a few beggars and what appeared to be local Cambodian tourists. As we went further in, and further up steps after steps, there was always more to see: fantastic carvings, dancing apsaras and Buddhas. We explored the monument in the early afternoon, having the entire complex to ourselves. Some monks led us into the central area.

Cows graze below Angkor Wat, Cambodia

Words cannot adequately describe the incredible carvings, reliefs and complex stone towers of the main temple. Built by Suryavarman II in the first part of 12th century, the temple honoured the Hindu god Vishnu. Much of its architecture bears similarities to smaller temples in India, Khajuraho in particular.

'Danger – Mines' signs by a village house

The tragedy of Cambodia was never far from the surface though. A few locals lined the main entrance, most of them very badly crippled by horrendous injuries received from hidden mines. One poor little girl had neither hands nor feet.

We continued our exploration, heading north through thick jungle with towering trees on both sides. Soldiers of the now united Cambodian army guarded the route, most heavily armed but looking fairly bored. Suddenly we came face to face with the great imposing face of Avalokiteshvara – a bodhisattva, disciple of the Buddha. These enormous faces (looking in four directions) sit high over a sharp arched gate and entice one into the great city of Angkor Thom. It is believed that up to one million people inhabited the 10sq.km. area enclosed by the five great gates and area now enclosed by overgrown walls some 5m high. Stone-carved gods and demons line the immediate entrance to the gates of Angkor Thom.

Perhaps the most stunning sights of Angkor Thom are the central towers of the Bayon with more than 170 sinister, half-smiling faces cut in stone with their glazed stares captured for the last 800 years. As we sped on from the Bayon along a broken road to the east gate of Angkor Thom, we could hardly have imagined the shock of the next few minutes.

For the journey back we thought of trying the long way round. While parked at the eastern gate listening to the early evening chorus of cicadas, there was a sudden deafening silence. A gunshot rang out, a solitary gunshot that shattered the silence and our nerves. In all the excitement it had been easy to forget the recent history of this truly fantastic monument. We hurriedly did an about-turn and scurried back to the guesthouse and another delicious calming meal served in a coconut. It was with some relief that the owner said it was not uncommon to hear shooting; probably bored members of the army shooting birds for target practice.

22 November
Up at 6am, we accidentally got twice as much breakfast as we ordered, but it was so delicious it didn't matter. Food may not be everything, but it certainly helps keep up your strength and morale. We were back on the bike just after 7am. This time we went east before Angkor Wat and the road was absolutely appalling! It is a trifle unnerving riding down these lonely jungle roads, knowing the Khmer Rouge are out there somewhere. A few buffalo carts lumbered into potholes and out of sandy courses. The overgrown temples of Banteay Krei and Ta Prohm have been left much as they were found before restoration began by the French in 1908. The vast area of Ta Prohm is particularly interesting, with vines and great tree roots strangling the temples. The dense undergrowth hides many fine Apsara carvings.

At Ta Prohm a friendly disabled man guided us through the jungle-infested ruins. For fun he shot an arrow from a crossbow into a distant tree with uncanny accuracy. He led us up an unlikely jumble of rocks on to the rooftops for a great view all around, then down another jumbled pile of boulders. We met no other tourists this morning at the temple and fortunately did not encounter the small deadly green Hanuman snake, which lurks in the dense growth around the temple.

At the furthest extent of the temple complex of Preah Khan we saw some graffiti showing a man shooting a UN helicopter with a large rifle or machine gun, and a man with a parachute coming down out of it. On the way out we met John Sanday, a British architect who is working on the conservation project there. He has also worked on projects in Kathmandu. He said he hadn't seen the graffiti before, and had heard rumours that Preah Khan was the next place to be attacked by the Khmer Rouge. I had thought of taking a photograph of the graffiti but felt distinctly uneasy at the thought of hanging around too long there, far away from other humans (or perhaps not as far as we thought!).

Finding the exquisite temple of Preah Neak Pean was very difficult, being accessed down a very narrow and lonely road overgrown with jungle. This intricate, pretty little temple is set in the centre of a small pond. We were completely on our own except for an elderly local man on a bicycle Again the apprehensions of being alone surfaced.

The sound of the buzzing/whistling cicadas in the late afternoon is truly overwhelming. At first we couldn't work out what it was – it sounded like a jet engine revving up, literally. We climbed a hill with a great view of the sunset over Angkor Wat, the Tonle Sap (lake) stretching out into the distance, and the hills which probably form the hideout of the Khmer Rouge. The soldier guarding it happily pocketed a dollar for the 'entry' fee.

Angkor Thom gates

There's just time for a quick wash before meeting John Sanday and his family for a drink and dinner at the Siemapheap restaurant. A great end to a great day at 10pm.

23 November
Not too early out of bed today, as we decided not to try to reach Banteay Srei. Too much hassle to fix a price, plus rumours about armed guards being necessary put us off. Instead we headed east about 15km to see the Roluos group. The countryside along the way was very interesting. Lots of wooden houses on stilts, lush green rice paddies, Bright red water lilies, children playing in the water, cows, buffaloes. An old lady with conical shady hat offered us a drink. No other tourists in sight as usual. The Bakong temple and adjoining structures are not set in jungle. At the temple a Canadian Red Cross clinic was taking place in the back of a truck: lots of visitors on bicycles.

We returned to the hotel for refreshments, then to our horror, just as we were leaving again, the engine cut out and couldn't be fixed. We were so lucky to break down at the guesthouse! Mr Proem gave us his new Super Cub 70 and we were off again, to the eastern side of the Grand Circuit and the elephants. We passed a tank and lots of army. In the elephant temple we met – again we were alone – a young soldier with a grenade strapped to his waist. There seemed to be more military presence around, perhaps because of the Prime Minister's visit today. Anyway we left after nervously taking photos of the elephants and retreated to the more well-trodden paths around the Bayon.

Exploring Angkor by motorbike

Angkor Bakong temple

This time we ventured down a dirt track to the west gate, in a somewhat dilapidated state. This gate had been left very much to itself in the ever-encroaching jungle, and was accessed by a poor, sandy track barely wide enough for the wheel of the motorcycle. At the end of the track we came upon a small shelter in the shadow of the great gate; the jungle was attempting to wipe the icy smile off the gigantic faces here. Along the pathway were the occasional short sticks of now disarmed mine tripwire supports. Bob went 50m or so further down the track to take a photo of a 'Danger – Mines' sign. My heart almost stopped as I saw two men coming up behind him carrying a bazooka over their shoulders – I was too far away to do or say anything. Fortunately they were from the army and just smiled as he returned and we drove back to the centre of the Bayon.

Here we found yet more carvings and murals to wonder at, and as we left for regrettably the last time, about 25 soldiers trooped out from somewhere inside. The Cambodian army has eyes everywhere in the jungle.

On the way back we stopped to take a last photograph of a 'Danger – Mines' sign. A local appeared from his shelter and grabbed the sign from its tree. 'How much pay?' for this sign, he asked. How many more signs did he have in his shelter, we wondered. Did this mean that the signs were now superfluous, or did Cambodia's new entrepreneurial spirit mean that mines could be ignored? The mysteries of Angkor are more than those of history.

That evening they showed a video at the Bayon restaurant next to the hotel. The audience, both local and foreign, sat absorbed, quiet and very still. Few people, locals or tourists alike, were dry-eyed at the end of this film. The film showing was The Killing Fields. This was a chilling reminder of the horrors of the Khmer Rouge, who are still now lurking in the hills around here. In a newspaper we just read that eleven people had been killed by the Khmer Rouge in a village south of the Tonle Sap. One of the saddest sights we have ever seen was a pretty little girl, about three years old, with no hands and no feet, sitting begging outside Angkor Wat.

Vietnam & Laos

After all the horrors of the American war on going in our youth, we were fascinated to visit Vietnam. Saigon or Ho Chi Minh City was a sweltering place of noise and vibrancy. We paused for thought outside the old American Embassy, from where the last remaining staff were helicoptered out as the Vietcong moved in.

The former French hill town of Dalat had a much more tranquil setting and wonderful coffee! Near the attractive coastal town of Nha Trang were the Cham dynasty ruins of Po Nagar and a beautiful bay that would soon be developed for tourism. The towns of Hue and Danang, names resonating from the war, were intriguing, with numerous historic forts, temples and the famous Marble Mountain.

Pagoda in Hue, Vietnam 1993

Hoi An fish market

A street in Hanoi, Vietnam 1993

The tiny fishing village of Hoi An showed a more traditional side of the coast. We really liked Hanoi, with its willowy streets, quiet ambience and the two Darling Cafés competing for new budget travellers' cash. A macabre sight was the mausoleum of Ho Chi Minh.

We flew from Hanoi to Laos and back to Bangkok. The trip proved an amazing insight into the turmoil of the post-Vietnam War period. There was an incredibly overwhelming sense of hope, a new breath of fresh air and vitality, a new beginning for Indo-China.

That Luang temple, Vientiane, Laos

Krabi beaches, Southern Thailand

The positives and the negatives

Twenty years on and tourism is now in full swing to Cambodia, with no danger to travellers or visitors. It seems strange that today's Gap Year students can travel so easily to these old war zones that we never expected to visit. Then again, we never expected to visit Tibet, China, Georgia or Central Asia.

Unfortunately today the number of safe places on the travellers' list is shrinking fast, as much of the Middle East and most of North Africa is lost to the new reality and insoluble chaos.

At this rate we will soon be warned not to travel beyond Europe!

The Pilgrims Way

The short man with a long, curly beard laughed out loud. It was an infectious laugh; no one could fail to hear it. But how come he was selling a huge pile of our route notes on a trek in Tibet, here in this bookshop in Kathmandu? We had only produced twelve copies, exclusively for Stanfords in London.

'Please Sahib, some tourist came in here selling this stuff. I didn't know he wasn't the author.' It was not convincing, but he offered us tea and charmed us with his stories. He promptly agreed to publish the booklets in a proper format and offered us money for the rights. In twenty minutes, it was a done deal. We have no regrets over this chance meeting. Rama Tiwari is the owner and publisher of many books under the Pilgrims brand name.

Rama Tiwari comes from a small village near Lucknow and Kanpur. He moved to Varanasi and sold books from his patch on the pavement, not far from the railway station. After some hard years, he moved to Nepal, where the scope for selling books to tourists far outweighed the scant pickings on the streets of Varanasi. The business prospered, and he met a lot of interesting people. Richard Josephson, an American living in Nepal, decided to help the budding business and soon a shop on the lakeside in Pokhara brought further rewards. Rama moved to Thamel and opened Pilgrims

Book House. The shop was quite small when we first bumped into him, but had a pleasant coffee/tea area.

Bob and Rama in Thamel, Kathmandu

A few years later in 1993, Rama agreed to publish our newest booklet, a guide to Mustang, which we had been fortunate enough to visit in October 1992 with our own small group. This was duly produced in both English and French, with colour pictures. After that we spent a lot of time helping in the publishing department, editing much of the prolific material that Rama uncovered in all corners of India – long-out-of-print gems, lost copies of historic works, books on the Himalaya from the Hindu Kush to Bhutan and Tibet. Some books, up to 200 years old, had long been unavailable in the west.

That was the beginning of a long and fruitful relationship that would last on-and-off for nearly 20 years from 1986 to 2009, when the publishing wound down in Kathmandu.

Sadly in 2013 the atmospheric Pilgrim bookshop burnt to the ground with the loss of many books, new and some very old. The new shop down towards Chetrapati is now run by two of Rama's daughters. Publishing is now confined to the Varanasi office, where Christopher, the expat British editor, works mostly 'flat out' – the only position to be in Varanasi during the hot season! He came overland in the seventies, reached Varanasi, married and never came back. He still hasn't been to Kathmandu.

Those overland buses again

Middle East: 1994

We were at a loose end in the UK again, doing temporary work in local factories at the beginning of 1994. It was quite a shock to be phoned by Geoff Hann asking if we could take a trip around the Middle East. The company had now become Hinterland Travel. Ashley Toft, who had previously worked for Explore, joined Geoff in this new venture.

Three in the morning at the old Athens airport – not a great start. Using various trains and endless local buses we eventually rolled up in Alexandroupolis near the Turkish border, where the bus was parked at a campsite. Even before reaching Istanbul we'd already had a puncture in the worn-out tyres – things were par for the course at Hann. The trip went very well; the Mercedes midibus behaved after a couple of initial teething problems. We visited some great places, including Palmyra in the Syrian Desert.

Palmyra, Syria 1994

Also on the list were Petra in Jordan, Wadi Rum and the Sinai in Egypt. Here St Catherine's Monastery and the climb up Mount Sinai were the truly high points. We collected a couple of trainees in Amman, so there were now more staff than clients. It was rather sad to leave the group at Erzerum in Eastern Turkey; we suddenly felt a great longing to do another full overland. But our season in the Alps was only a few days off, so we had to return to the UK, with new ideas in our heads and a dormant bug ignited.

Old waterwheel in Hama, Syria

Gethsemane, Palestine 1994

Asia Overland trials: 1995–96

In the spring of 1995, with less winter work on the horizon, we went tramping in New Zealand to develop some new supported trekking ideas. It was also a good opportunity to catch up with Siân's schoolfriend Sue Lawrence.

In autumn 1995 there was no trekking work in Kathmandu, but some editing to do at Pilgrims. An exciting request came from Geoff Hann: could we take a few clients on a trip to Rajasthan and Gujarat? I started fixing up the same white bus we'd used in the Middle East. It was not such a pretty sight now and had suffered in the meantime.

The customs papers were also not in order. Somehow the time the vehicle had spent in both India and Nepal was adrift with the rules. Rules were after all bent frequently, but problems were looming.

The trip to Gujarat proved fascinating; neither of us had been before. The city of Bhuj, later destroyed by an earthquake, was a great, atmospheric place. Junagadh had some amazing architecture; almost Disney-like, and the rare Asian Lions came out to play for us in Gir National Park.

A welcome in Gujarat

An extra passenger
Another passenger, unwanted, was a rat, which chewed its way through the metal lid of a pot of jam and more importantly found a home behind the dashboard, where it proceeded to chew its way through the electrical system. Something had to be done, but we didn't want to kill it. So we tried devious means such as hanging a piece of cheese on a piece of string outside the door overnight, hoping that it would fall out of the door and be unable to climb back in. How naïve! Soon the cheese was gone but bits of plastic wire sheaths, along with nests of music cassette tapes, continued to appear. Sadly a search of the local shops for rat poison was the only option and the following morning we saw its bloated body lying in a muddy puddle below the door.

Curious mosque in Junagadh, India

Bijapur palace

We got completely hemmed in down a narrow road that became progressively narrower as it led into a bazaar in Rajkot. The locals were not initially best pleased, but eventually some helped to clear a way as we reversed backwards for more than half a mile. In Central India, some of the highlights were Chittoor, Ellora, Ajanta and Bijapur, with palaces, atmospheric streets and the vast

Gombaz mausoleum. The clients were rather difficult at times on the trip and we were pleased to be finished with them in Goa! Here a new client joined us for the run back to Kathmandu.

The cow, an enormous Brahmin, climbed the stairs and stepped through the door, sniffing its way into the restaurant. At the far end it raised its nose and attempted to grab some food crumbs. The waiter turned nonchalantly and paid it little attention; perhaps this sort of thing happened every day. We were in Kolhapur, sitting in a dimly lit restaurant eating a great selection of dishes, stuffed tomatoes, palak paneer (spinach with cheese). Eventually the waiter grabbed it by the horns and turned it around; the cow was now tired of the proceedings and retreated as unconcerned as it had first entered.

The route to Nepal took us through Mandu again and then to Jhansi, where we found the amazing old colonial hotel, the Jhansi Hotel. It had barely been touched since the Raj departed; it might have been cleaned in the meantime. The TVs worked but not the plumbing. It still needed a man to go out the back to light a fire for hot water, but the man was not to be found that day. The food was superb though, the waiters all Nepali.

The tranquil Mandu Palace, India

Customs problems with our bus now intervened, causing mayhem; it was allowed only 6 months in any one year in Nepal. Clients were arriving after the bus was due to be out of Nepal. The local Nepali agent for Hinterland arranged for the clients to be bussed down to the border, where we were sorting out the problems. The Indians then refused to put the bus on to the new carnet (customs document). Now we were forced to leave India early as well. Things went well as far as Delhi, but the back axle developed an ominous problem, with far too much movement in the pinion bearings. There was no time to fix it in Delhi and the customs again refused to be helpful. Clearly a previous driver had bent the rules just too far this time.

We spent over a week in Lahore fixing the problems, as parts were hard to find. The town was in the grip of cricket mania, the World Series being played up the road. It was all great fun, an insight into a different side of Pakistan and refreshingly down to earth. Mrs Thatcher also turned up in Lahore to give some lectures, staying at the Avari Hotel. We used the hotel for faxing London; its afternoon tea buffet was great value even for those on a budget.

Some of the clients were running out of time by now and it didn't help that when we finally got going, a new problem developed in the front brake. We never did get north to Peshawar and Rawalpindi. In fact we made it no further north than a town called Gujranwala. Gujranwala is one of those great masses of seething humanity, choked by all manner of traffic, mechanical and otherwise. Its streets are heaving with pollution and it's none too clean – one of those places you either love or hate, rather like Gorakhpur in India. Gorakhpur has Bobi's restaurant though and it's the last place in India before the peace and tranquillity of Nepal.

The Fort area, Lahore, Pakistan

Heading south and west now, just out of Multan the bus finally gave up the ghost; the engine refused to go any more without a large injection of money. The clients were by now not happy punters and some decided to make their own way on to Iran. We were forced to deposit the bus with the Pakistan customs in Multan: it was given two months in which to be fixed or surrendered to them. I never heard the outcome.

Siân and I made haste back to Kathmandu to collect another bus. In Amritsar we had a relaxing and memorable evening watching BBC World. The police drama A Touch of Frost was showing. The room was a haven of tranquillity, a sanctuary away from the realities of life in India. Next day we began a gruelling train ride to Gorakhpur and a night bus to the Nepalese capital.

The replacement bus was not in a great state either, but it was fundamentally much newer. In order to save time we decided to take a shorter route to Delhi on the newly opened road through western Nepal. Unfortunately some of the bridges had not been connected to the road and we took the bus through some harrowing riverbeds. Back in Lahore we collected the one remaining client and headed on to Quetta. In Esfahan we finally caught up with the rest of the clients; they had hopped on to an Exodus truck that was being driven empty to the Middle East. Since our bus needed some small repairs to the front axle, we still couldn't leave with them.

It all ended rather sadly with us driving to Istanbul on our own. It was desperately cold and not a good end to the trip. It did, however, galvanise our thoughts; we decided to run our own overland in the future.

Chaos in Multan, Pakistan

Luckily we found more trekking work in Nepal through our long-time acquaintances, the Millers. They were running Explore's Nepal program from Kathmandu. The programme of treks was intense for such oldies as us, with back-to-back treks to Annapurna and Everest for most of the season. As always, the Kathmandu merry-go-round was graced by many great characters, both local and foreign.

In-between treks, work continued at Pilgrims, where a never-ending stream of books needed our attention. It was great to be able to get into the mountains on trek and then spend time in Thamel working on some fascinating books. Such titles, although very specialist, gave us a wealth of knowledge, with subjects as diverse as faith healing, Tantric Buddhist philosophy, historical journeys through all of Asia and some other very esoteric subjects.

Siân and KC at Manakamana cablecar 1998

KC – one of Kathmandu's colourful characters

K.C.'s story started with humble beginnings and limited schooling. But he had a vivid imagination and a strong desire to travel and see the world. From his early days as a waiter in the hotel Soaltee Oberoi, he learned what foreigners wanted – 'Would you like sugar, milk, sir?' - and had soon gained the contacts that would help him to get a job working for, of all things, the Royal Nepal Shipping Corporation. Flying from Kathmandu to Calcutta and then on to Karachi with Japan Air Lines, he then sailed via Cape Town to London with a cargo of jute.

Gradually he got to know foreigners' eating habits and gained much experience in dealing with their appetites. It was his restaurant that really started the growth of European-style food in Thamel. K.C. himself was always there, with his long flowing hair and a cigarette in his mouth. He looked more like a hippie than the real hippies! His was the very first 'sizzling steak' and salad. People would see the clouds of steam go sizzling by and say 'Mmmm... What's that? I'll try one of those...' His was also the very first cheesecake, made with Kraft cheese.

K.C.'s first began in Pig Alley near Freak Street; buff steak was then three rupees, fifty paisa at his place called Bag End. He was one of the first to move up to Thamel. In the early days there was no furniture – people would sit on the floor in traditional Nepali style with cushions. But then he moved on to having seating for sixteen, and gradually the restaurant evolved with the famous sizzling steaks and fabulous cheesecakes. The place was always busting to capacity with overlanders, trekkers and odd ball-like us!

And so what does K.C. stand for? Is it Kaput Crazy, as he used to say, or King of Cuisine?

****

In the yard near Winchester were three Swansea city buses. None of them looked wonderful, but at least they came cheap. We chose what appeared to be the best of the bunch and handed over £2500. During the winters of 1997 and 1998 we worked flat out on the bus, repairing or replacing everything from end to end and top to bottom.

Cuba, Dominican Republic and Haiti: 1997–8

With all the work we had to do on the bus, we had only short trips away, visiting Cuba, Haiti and the Dominican Republic.

No one can fail to be drawn to the amazing 'experiment' that Castro enforced on Cuba for the last few decades. Its health service is a major achievement and the relaxed nature of life seems another plus. Delving deeper under the exterior visible to visitors is much more difficult. The obvious attractions are there; the quirky sights of Havana, old American cars and faded Spanish colonial character, the city of Cienfugos, the tranquillity of Trinidad and the beautiful landscapes of Vinales.

How will Cuba change now that it has relations with America?

Havana street scene, Cuba

While Cuba was relaxing and intriguing, Haiti was not. We took a bus from Santo Domingo in the Dominican Republic and arrived during the Carnival period. This in itself was a great bonus, except that the crowds proved unruly and surging mobs made the experience very scary. That was apart from the reputation of the country as a wild and crime-infested domain. In fact it wasn't as bad as portrayed and the sight of Sunday morning mass in the great cathedral of Port-au-Prince was worth the visit in itself. Sadly it crumbled in the subsequent earthquake of 2010.

As for the Dominican Republic end of the island, it too seemed to have a fair number of unsavoury characters to be avoided. It was easy to see why the package holidaymakers on the island avoided most of the country, staying in their safe hotel compounds. Yet as always there were some helpful and friendly people to engage with and a plethora of sights in the old area of the capital.

Cathedral in Port-au-Prince. Haiti

Santo Domingo street

169

PART THREE: THE NEW MILLENNIUM

Overland for Oldies, No Problem: October 1999

On a cool evening in October 1999 we drove to Portsmouth to begin our own ten-week expedition across Asia. None of our clients was less than fifty years old and all had an amazingly adventurous outlook. We already had a photograph of one of the two passengers we were collecting here, so he was easy to trace. But I wandered around the shops and restaurant for ages trying to pick out a likely woman. Who could it be? I met David and we continued our search. Eventually we asked the desk to page her for us. 'Will Susan Wright please come to the information desk.' The woman next to us turned round. She was pushing a trolley with two enormous red holdalls and two other smaller bags. Our dossier did mention there was no baggage limit, assuming that people would bring only a reasonable amount! But who is to decide what is reasonable? Susan turned out to be a good friend with whom we still keep in touch, but she was not known as the bag lady for nothing! At 11pm we left the harbour, gazing out at the lights of Portsmouth as they faded into the distance. Now it was really happening; twenty-five years since Bob had first done the trip, a Swansea City bus was now on its way to Kathmandu.

The next morning dawned foggy, and we had driven for hours across northern rural France before the skies cleared. The night was spent in a delightful hotel, like a mediaeval country house, in a secluded town north of Beaune. We crossed into the familiar territory of the Swiss Alps, Martigny in the Rhone valley, and stayed half way up the Col de Forclaz at Ravoire. The hillsides were ablaze with golden autumn shades, particularly at sunset. The next day was Pisa, the next Sienna, Assisi and the mediaeval port of Bari, on our whistle-stop tour across Europe.

The super Superfast ferry whizzed us across the seas to Greece. Unfortunately the Superfast ferry going the other way caught fire during the night; no one was killed but some of the vehicles on board were destroyed. In Greece we drove via Ioannina, Metsovon and the monasteries of Meteora before entering Turkey.

Our bus at Ihlara, Turkey

Turkey

The rest of the group flew in and we nudged our way through the choking traffic of the city to cross the Bosphorus into Asia. Istanbul had grown dramatically over the years. Just a few months before a terrible earthquake had struck the region of Turkey close to Ankara; we saw masses of tents and temporary housing. In fact another earthquake struck just a couple of days after we had left the area, with some terrible landslides on the road we'd taken.

It was freezing at Goreme in Cappadocia and even the novelty of sleeping in a cave hotel did not keep us completely warm. The hotel had been constructed in the soft tufa volcanic rock, giving it a cosy atmosphere. Once into eastern Turkey the going was slower, with three very high passes on the way to Erzerum. Winter was fast approaching by now and storm clouds obscured the view of Mount Ararat.

Iran

The Iran border was nothing like as intimidating as in the years just after the revolution. A man from the tourist office helped us to sort through the endless paperwork, while also helping himself to a nice profit by changing money for the group. He then invited some of the group to visit his house in Tabriz; they found it a fascinating insight into life in the Islamic Republic, his official public persona being in marked contrast to the relaxed, informal atmosphere in his home.

Driving on the fast open roads across the desert, while enveloped from top to toe in black robes, was quite a challenge for Siân!

Everyone loves Esfahan, the majestic showcase of the Persian Empire with the timeless visions of the great blue-domed mosques. We also visited the Armenian area of Jolfa for the first time, being surprised by the intricate and strong style of the cathedral. After exploring the old city of Yazd, our merry throng reached Bam. In Bam we stayed at a new barely-open plush hotel before the rigours of the desert of Baluchistan. Over the years a lot of work had been done restoring the ancient citadel of Bam, but in 2003 a powerful earthquake reduced much of it to ruins.

Rumours that we would be required to have an escort from Kerman to the Pakistan border proved true. It was said that this was to protect us from possible kidnappings by drug smugglers and bandits operating across the nearby Afghan border. A tour group of Italians had been kidnapped a few months earlier from the same hotel we used in Kerman. It's a pity, because the hotel is one of the most efficient, pleasant and accommodating in all of Iran. They pride themselves on their excellent home-style Persian food, which cannot be found anywhere in public restaurants. The normal fare is butter rice and kebabs.

Caravanserai in northern Iran

Citadel at Bam, Iran, before the earthquake

Escorted from Iran to Pakistan

Our first escort proved rather nerve-wracking; first of all we couldn't keep up with them and then we were taken to a police station, the passports taken and a long wait ensued. The ice was broken slightly when some of the group needed to use the loo behind the steel doors of this compound. The next escort proved much more relaxed. For some reason no one escorted us to the border town of Zahedan, about as close to the Afghan border as you can get. An Encounter Overland truck was also escort-less. They seemed very surprised to see our coach, and even more surprised to see it full of grey hair.

Pakistan

Once in Pakistan, things as always seemed much more relaxed and friendly. The customs officials brought us tea and sweets, told us we could remove our headscarves, and took only a few minutes to do their work. A new road opened before us, a super smooth road in place of the ghastly rock-n-roll of the previous dirt track.

A new hotel in Dalbandin provided a better night's rest than the old camping bungalow ever did, even if power cuts still featured most of the night. After Dalbandin is a short section of small sand dunes that everyone wants to photograph, then the road up to Quetta is somewhat forbidding as it runs close to the Afghan border.

Dark mountains and sharp rocky passes haunt the route. Quetta is like an oasis in this barren, inhospitable tribal region. Unfortunately the hotel had declined badly with the dearth of tourists in recent years. Only the Farah restaurant provided the same quality steak, onions and chips, the waiters as always providing some entertainment.

Descending from the Bolan Pass, the landscape changes dramatically – from sinister, barren mountains where whole armies have hidden in the past, to dusty open plains. Another new hotel, this time in Jacobabad, saved a sleepless night in the old Dak Bungalow, where Benazir Bhutto's entourage had taken our reservations on a previous trip. It was a pity about the very loud music for half the night, and astonishing numbers of fierce flies.

On the road to Quetta, Pakistan

On the highway between Karachi and Lahore, the traffic moved slower than the flies. Road building created clouds of dust. Even the colourful old trucks ground to a halt in all the chaos. The road, despite its dual carriageway status, was in places like two separate roads side by-side; no one kept to the correct side of the road. A dual road to the death indeed!!

On the road from Jacobabad

Sufi shrine at Uch Sharif

The Sufi shrines at Uch are a reminder that Islam has many facets. Sufis today are often much maligned by hardline Muslims for their more liberal themes, including song, dance and poetry. More Sufi monuments and mausoleums are found in Multan, en route to Lahore.

Lahore – a great melting pot, thankfully no cricket matches this time to distract the crowds, the drivers and the buffaloes. At Faletti's, a once-majestic palatial hotel of the Raj, the melancholy of the lack of tourists showed in the great swathes of peeling plaster, rusty pipes and dribbling taps. Only the staff, the doorman and the waiters still hinted at the pretence of the luxury of a time gone by. The place was quite literally wreaking of atmosphere; one hardly dared to open a window for fear it might fall out or disturb the inward tranquillity by exposing one to the noise, fumes and confusion of the streets outside.

India

Wagha Road, or is it Attari Road; it's all the same, but one is the Pakistan side and one the Indian. Here is a pageant the likes of which one could travel thousands of mile to behold. In this carnival the atmosphere is serene, everyone knows his routine, traditions are kept with absolute faith. No one acts with any sense of fear or apparent hostility. It's a game; a spectacle with few spectators but it never stops. From dawn to dusk the ceremonial exchange takes place in all weathers and under a burning sky. The porters in their amazing and colourful uniforms exchange their wares, a ritual handing over across the line. At close of day is an even more exotic ritual, viewed by tourists from around the world.

Siân & Sikh attendants, Golden Temple, Amritsar

It was good to be back in Amritsar. The Sikhs are so hospitable and such efficient mechanics too. Things are more ordered, although to a western mind it's only relative compared with the rest of the subcontinent. The Golden Temple complex was almost all restored. Restoration of the Akal Takt building was nearly completed, except for the intricate painting work high up on the inside ceilings. This building had been all but obliterated by the 1984 army invasion to clear the temple complex of the militants and other followers of Jarnail Singh Bhindranwale.

But now turbaned men of the most elegant stature surrounded us; all were without exception most courteous, polite and pleased to welcome us.

We stayed at that most traditional of overland establishments, Mrs Bhandari's, in her efficient guesthouse. Located in the former British military cantonment area, the house and garden are as charming as a house in leafy West Sussex. A Swiss coach was parked in the yard with some younger overlanders.

The Line of Control

A normal country border exists between Lahore and Amritsar, but further north the exact line of the border is in dispute and is referred to as the Line of Control. This line marks the zone in which a nuclear exchange between Pakistan and India could be the worst outcome.

At the Wagha/Attari border, the guardians of this line go about their daily routine, exchange pleasantries and appear to care little for their masters in Islamabad and Delhi. It is all rhetoric to them, although they actively put the other side down whenever possible. It seems like a charade. Sense and reality are teetering on the edge of an abyss, an absurdity under any other conditions but for the carnage it might invoke.

Half of the group departed from Delhi. On the last night we were invited to attend the Christmas Carol concert in the garden of our hotel, the YMCA, a little of old England in the heart of New Delhi. The departing clients were getting late for their flight, so we bundled them into a taxi. The driver was a Sikh, of course, the taxi an Ambassador, the Morris Oxford lookalike. It was a sad farewell; it had been an interesting trip with some amazingly congenial clients.

At the gates of Fatehpur Sikri, India

Next morning the rest of us were on the road again, bound for the Pink City of Jaipur and another round of sightseeing with Mr Pram. It was a long time since we had last taken a group to Jaipur, but nothing much had changed. The narrow alley to the hotel was no wider, the rooms were almost but still not quite clean and the tea was still so strong that the leaves almost forced the lid off the teapot. Only the crowds were bigger. We took a detour to Deeg; it was mentioned in the Lonely Planet guidebook as a little gem, so it must be good. It was actually pretty good, with its massive gardens, city walls and palaces, but it wasn't quite as good as its write-up and the road to it from Bharatpur was certainly dreadful.

Jhansi, Orcha, Khajuraho, all these old places and old faces... but now we were in our own bus and with our own clients. What a worry, but what a trip. Satna, Rewa, Mirzapur... that acrid, smelly den of iniquity famed only for its narrow bus-eating bazaar. Famed also for its dacoits – bandits and hoodlums – goondahs of ill repute. Then we reached Varanasi, so we had to have some mechanical breakdown for Mr Shafiq to fix. The brakes were a bit knackered since the terrible road near Satna. It was the only mechanical niggle in nearly 8000 miles.

New Millennium Nepal

Christmas Day found us in Chitwan National Park for a feast of wildlife and entertaining Tharu stick-dancing. At Pokhara were some welcoming familiar faces at the Tibetan-run hotel. For years we had dreamed of driving up the winding hilly Rajpath into Kathmandu. Now it seemed almost an anticlimax as we parked up for the last time. A new millennium was about to dawn. On the last night we celebrated at the Rum Doodle restaurant, a fitting end to a unique trip.

Our bus in Bhaktapur, Nepal

Interlude in Bhutan: January 2000

Being in the Nepal Himalaya for much of the year, we had long dreamed of the chance to visit Bhutan. It finally happened in 2000, with a brief trip to Paro and Thimphu. No one could fail to be overawed by the massive Tashi Chodzong complex in Thimphu, but for the best picture-postcard scenes, Paro Dzong, Tiger's Nest monastery and the Paro valley take the top slots.

Paro Dzong, Bhutan

The problems of Shangri-La
Although a beautiful, tranquil place with some impressive mountains and monasteries, there is a slightly disarming feeling about the way it's not all that it seems. Thousands of Bhutanese-born Nepalese were sent into exile in refugee camps in Nepal, in a kind of 'cleansing' of the country. This situation is still open for discussion, as many Nepalese were also immigrants. The country does seem to have more positive vibes than negative, though, with happiness being a national goal.

Overland back to the UK: February 2000

One last surprise awaited us. After twenty-six years of avoiding a 'proper job' we finally had the great pleasure of being joined for the Kathmandu–Delhi section of the return journey by Bob's parents, Tony and Beryl, from pig breeder and egg lady to travellers on the hippy trail. Hippies without a flower in sight, a new age experience never contemplated in so many years for lack of opportunity or means.

Together with Lynne, another journey began. With the wheels off in Pokhara, it was little different from being on the farm, another job for father to help with! There were the usual mosquito wars in Sonauli, Mr Chapati's tour of old Varanasi, silk sellers and Shafiq meeting the folks. A mini Kumbh Mela was in full swing near Allahabad; we got lost in the bazaar once again. Tony and Siân hired a rickshaw to guide the bus to a hotel in Kanpur after fruitless endeavours to find one that had a carpark. Tony, who was too old to have taken a UK driving test, just couldn't believe the Indian traffic rules.

We hired bicycles to visit Bharatpur bird sanctuary, oldies as well. Butter Chicken was on the menu at Bharatpur – it curried no favour with Tony. Eugene Pram guided us in Jaipur and everyone ate like crazy in Delhi. This trip was little different to any of the others, but a rather sentimental journey for Bob with his parents, who flew back from Delhi.

Seven weeks later we rolled into the farm near Chichester after a pretty trouble-free run. The only problem had been the injector pump, which had needed a new union valve near Marand in north-west Iran.

It was sadly the last time the bus managed to go Asia Overland, with the effects of 11 September casting a long shadow over Pakistan in particular. And now the Maoist rebellion in Nepal threatens to overrun the happy valley.

****

We were soon back in the Alps for another summer of work for Sherpa Expeditions, trekking with groups in Switzerland and also the Dolomites in northern Italy.

The Land Rover with a camping group in Lauterbrunnen, Switzerland

above Zinal, Switzerland

Later in Nepal the Maoist rebellion rumbled on in the countryside, but we saw little trouble in Kathmandu during the winter season.

The Dark Age of Kali: 2000–1

The jungle below almost wipes the undercarriage, as the plane comes in low over the hills of the Mahabharat. Picture-book villages cling to the steep-sided slopes; the rice is ripening for the harvest. A crescendo of colour from yellow to green spreads up the hillsides. The mighty jet veers from its normal path and circles around to the north. Banking to the left, the wing tips appear to brush the shrine below. The eyes are watching, the flags fluttering in the breeze, as Boudhanath smiles at the passing metal bird as it levels up and dives down for the final approach. Kathmandu, we are back, but what does the winter hold now that the King has been slaughtered and the Royal Family all but wiped out?

Another season in the Alps is over; there is no trekking work, but Pilgrims have a new project. There is a whiff of foreboding in the air. Maoist rebels have tightened their grip on the countryside. There are stories of intimidation, fear and brutality. Is this the same Nepal we have known for so long? Even the trekking routes have not been spared. Everest is increasingly only safe north of Lukla; the route through Jiri no longer Maoist-free. The airfield at Phaphlu has been attacked. From around the Annapurnas come stories of robbery and intimidation of tourists and groups.

The previous season we ourselves had witnessed the general breaking down of law and order. First trouble broke out over some alleged comments by an India film star, Hritik Roshan; a child was killed in the ensuing riots and tear gas was used outside the Malla Hotel.

On several days there was a general bandha – general strike – in Kathmandu. With no traffic allowed, people were forced to travel to and from the airport in cycle rickshaws.

A Bhutan group, including Siân's parents

Fortunately when our group arrived it was a normal day. We enjoyed New Year's Eve Millennium dinner in the Rum Doodle restaurant, but just after we had paid our bill there was a loud explosion behind the bar. Everyone jumped out of their skins. It was a tense moment, especially with the current political uncertainties. But a moment later we realised it had just been a gas explosion – one of the helium balloons had landed on a gas cylinder and exploded into a burst of golden flame. Discretion being the better part of valour, we left and returned to the Kathmandu Guest House lobby, where we watched proceedings on the television and celebrated the New Millennium for the second time, by linking hands with a group of Japanese singing Auld Lang Syne in the courtyard!

The next day, New Year's Day 2001, all of Kathmandu was once again at a standstill. We crept tentatively out of the KGH to check out the situation in Thamel. K.C.'s was open, though a young waiter stood guard at the door, ready to pull down the iron shutters at the slightest sign of trouble. We sat inside and savoured the taste of bacon sizzlers with filter coffee, unperturbed by the happenings, or lack of happenings, outside. Later that day we walked carefully down the old streets. Nothing much was going on. Most shops were shut, but the streets were full of children and adults playing football. It's not often that you can move freely in the narrow byways and breathe the air; in some ways, the day was a breath of fresh air, though of course the political situation was to get much worse in the future.

A day late, our small group was finally bound for Darjeeling and Bhutan. We visited the Rama Sita temple in Janakpur, a southern city close to the Indian border and famous for its artwork. But the next day was to be a long one.

Driving along the Terai plains, we came across the first barricade – a burning tyre and just a few young men. We persuaded them to let us past, explaining that we were just going to India and would not trouble them. But we didn't get much further. With more barriers across the road ahead, an angry mob and police riot squads patrolling nearby, we were going nowhere. The police invited us to wait in the grounds of the police station, a small garden with beautiful temple. It was peaceful enough, but there was no food or refreshment nearby – how fortunate that we had supplies of homemade Christmas cake to keep us going! An eerie calm pervaded the air, nothing stirred; all we could do was wait for the mobs to disperse. To calm their nerves, John and Marianne played cards with Gwyneth and Alan.

Behind us another barrier barred the way. At lunchtime some of the riot police arrived for their meal of dal bhat across the road in a café of dubious distinction. After their meal, some of them came over to watch the card games; most of the lads were barely out of school. They soon caught the hang of the game; it was perhaps more stimulating for them than the tasks that lay ahead. 'Pick up two!' they would call out whenever they saw a '2' in someone's hand. 'When will it be over?' we asked. 'Maybe five o'clock, maybe six o'clock,' was the reply. At nine o'clock we gave up and were heading back towards Lahan, where we hoped to find a hotel.

It was dark and an unruly crowd had gathered around the single tyre that had been burning when we had passed earlier. A mass of dark faces came ominously and frighteningly towards us. We had no escape as they surrounded our tiny vehicle. But this is Nepal, a land of peace and tranquillity. 'Don't worry, they only want to see who we are; after all, no one has passed them since this morning so they must be puzzled,' we reassured our passengers. 'No passing, mister,' they cried. Bob opened the window and said, 'Namaste, good evening, we are old people, you know, and guests in your country. Please help us to get to a hotel.' It seemed to be enough to get them to clear a path for the vehicle. Beyond this point on the road thousands of people were now stuck waiting, their buses parked along the roadside.

We retreated to a dingy but very welcoming hotel for curry. Outside a cacophony of horns began blaring out. The buses were finally moving! But we had already decided to wait until the next morning; our drivers were due to return at 5am. But at 4am there was a knock on the door. 'Let's go now, Didi, before the rebels get up…' Within minutes we were on our way to the border for breakfast and safe passage to India.

En route to Darjeeling, in Siliguri we visited the Hotel Mainak, which we had failed to reach yesterday. They were most helpful and we had a sumptuous lunch. 'It's lucky you didn't come yesterday,' they said. 'There was a strike here in Siliguri.' Apart from the obvious old colonial relics of Darjeeling, there are other attractions such as the Tenzing Norgay Mountain Museum, the Snow Leopard breeding centre and the Zoo hosting Red Panda. The 'relics' included the Planters' Club, where we had tea surrounded by stuffed tigers, bears and yaks.

Bhutan was not on strike, and the trip was a great success. As a bonus, the Tiger's Nest Monastery outside Paro was accessible after a terrible fire a few years before. The trek up is a very spectacular walk, climbing to a high viewpoint of the monastery. This time around our route took us to the old capital of Bhutan, Punakha Dzong. The vast structure of the fortified monastery is very imposing, sitting between two turquoise rivers. A little further along the road is Wangde Phodrang, another of the country's fine legacies of the past.

Darjeeling Tea

Punakha Dzong, Bhutan

The flight from Paro to Kathmandu must surely be one of the most incredible in the world. The peaks of Chomolhari, Kanchenjunga, Makalu, Everest, Lhotse and Gauri Shankar are the stars on the northern horizon.

Militant tendencies: 2001–2

Let us now move on to the winter of 2001–2, post 11 September. Our overland trip has been cancelled, because of the war in Afghanistan and increasing insecurity in Pakistan, with militants targeting foreigners.

> ### The Maoist Rebellion
> In late November 2001 a state of emergency is declared, the Maoists have attacked and murdered over 150 policemen and targeted the army for the first time. The streets empty of tourists almost straight away; the Kathmandu Guest House, where we always stay, for the first time ever is operating at only two-thirds of normal occupancy. Even during the revolution of 1990 it remained full. Talks between the Maoists and the government have broken down and the rebel leaders have fled into the hills. The western part of the country has effectively been taken over by the rebels, the region around Dang being particularly troubled. In the city things remain essentially peaceful. A Coca Cola factory has been bombed, but no one was injured. Tourists have not been targeted.

In Thamel at the Pilgrims office, Rama has arrived from Delhi and Varanasi. There are already fewer tourists in the streets and the bookshop is not so busy. Our new project is to write some introductory books on some very stimulating subjects. Subjects that we don't have much knowledge of: the Kama Sutra, Tantric Buddhism and the erotic art of the Kathmandu Valley. This last subject offers the prospect of some very interesting research. Rama wants a lot of photographs from around the valley, of temples, the art forms and any other interesting curiosity we might find. We are looking for the green-eyed yellow idol for real now, not just as a dream for the future.

The naughtiness of Nepal as art forms, as trek leader Ann Sainsbury so aptly put it, can be found on many, but not all, of the temples in the Kathmandu Valley. The erotic images are mainly found on Shiva temples dedicated to the goddess of destruction, Kali and Durga. No one knows exactly why such imagery is found on essentially religious places, but many reasons have been put forward. The most plausible idea stems from the fear of the destructive gods, for they alone can destroy and give rebirth.

The Black (Kalo) Bhairab of Kathmandu

> **Interlude in India**
> The helicopter flies low, swooping to drop its cargo over the crowds. Its flight path cuts through the silence with a tremendous whirl of air, like a Maharajah's sword. Its cargo is dropped on to the expectant crowds below. A million butter lamps flicker in the reddening glow of sunset. The river water laps against the banks, where pilgrims clamour to see what the helicopter is dropping. A million drops of incense and rose petals descend to the mighty river; the ghats are alive with celebrating Hindus. The Karthik festival in Varanasi is a spectacle not to be missed. It was lucky to see this colourful celebration before returning to Kathmandu. We have been to visit the bookshop and publishing centre of Pilgrims in Varanasi. Rama is in the long slow process of setting up a printing press in Varanasi. Business in Nepal is threatened and he needs new avenues to preserve the company. The wheels of bureaucracy in India turn very slowly, slower than a farmer on an ox-cart trundling home along a potholed road at the end of a hard day's work in the fields.

Our work goes on in the serenity of the Pilgrims office and garden restaurant. Life is OK for us, a paradox in these difficult times. We are enjoying the new challenges, learning more about Hinduism, Tantra, Buddhism and Shamanism. Hom, the office manager, is a modern computer whizz-kid, one of the new generation of well-educated Nepalis. He has a different project on the go, sorting through a new book written by Naresh Subha, who worked for Pilgrims for several years. We are required to assist with this book on the different ethnic peoples of Nepal. Miss Rhicha returns to work at the office in late November. Her family live in the eastern part of the Terai, near the border with India and Darjeeling. She has returned on the night bus, a terrible journey of around 16 hours. It is now a journey of some trepidation, for Maoists have been exploiting the remote areas along the border. We are learning a lot this season.

The traditional inhabitants of the Kathmandu Valley, the Newari, are craftsmen and traders, priests and politicians. Some are Hindu and others Buddhist; in consequence they celebrate many festivals almost continuously throughout the year. In October the great festival of Dashain takes up much of the month, with offerings to Durga, the destructive goddess. Tihar is a more relaxed festival, with lights and firecrackers, revering dogs, cows, crows and brothers. In the spring comes the dye-throwing festival, Holi, after the Buddhist New Year, celebrated at Boudhanath. At Pashupatinath during the nights of Shivaratri, thousands of devotees gather to pay homage to Lord Shiva, the most important god in the valley. The ferocious idol of the black Bhairab, a ghastly image that observes the scenes in the Kathmandu Durbar Square is worshipped. Just once a year the even more hideous idol of the White Bhairab is displayed. These gods of destruction must be appeased if the future is to be positive. This winter (2001) a new form of rebellion takes a grip of the country, attempting to sweep away the vestiges of the old ways.

The New Nepal 2001
Nepal is coming to a new age; gone are the quiet streets of tumble-down houses, gone are the peaceful fields of the valley, long gone are the hippies. Gone too are the holy cows from the empty streets; even the barking dogs have fled in despair. In the countryside the farmers are being overwhelmed by insurgency. New ideology, social inequality, poverty and political frustration are the new bywords. A younger more aggressive population wants progress, hope and development. The democratic experiment is faltering along the way, as it takes shape and attempts to grow. It is unlikely that any answers will be found in the fallen ideology of Mao. The monarchy is under threat and the new king has a delicate balancing act to perform. There was never a more defining moment for Nepal.

Annapurna II, a timeless landscape, Nepal

The Caucasus unlocked: 2001

28 March: Baku
Landing at 3am in the morning at a strange airport with stories about dodgy taxi drivers did little to calm the nerves. Soon enough though we were speeding along a fast and fairly smooth dual carriageway, our jolly driver pointed out some sights along the way and delivered us safely to the Apsheron Hotel. We were on the 12th floor, with a fantastic view over the Caspian Sea – to be seen tomorrow!

At 5am we collapsed into bed …

29 March: Baku
… until 12:30pm, when we thought we really ought to get up and see the city!

Leaving the high-rise, chunky Soviet hotel, the sky was grey but it was warm. On each floor was a usually plump lady probably 'guarding' the customers. Years later this iconic but ugly building was knocked down for far more glitzy towers befitting a modern oil state.

In the old city we stumbled upon the Karvansaray Restaurant and settled into a cosy arched room for our breakfast/lunch at 2pm! It was a very atmospheric place, all the old merchants' rooms around the central courtyard have been made into individual dining rooms, with gas fires like imitation wood fires. Beautiful carpets decorated the walls and there was very attentive service. The meal consisted of chicken kebab, vegetable kebabs, a plateful of stuffed vine leaves, a bowl of yoghurt, two huge breads and an enormous pot of tea; and all this for £4.

The hotel in Baku, Azerbaijan

Suitably refreshed and revitalised, we then wandered past the circular and strange looking Maiden's Tower. Nearly 30m high and with walls around 4–5m thick, the building has a long history dating back to the 7th century B.C. Nearby is the old market square full of old stone carvings. More colourful carpets were on offer along the narrow lanes and cobbled streets with beautiful architecture all around. On the other side of the old city is the 15th-century Shervan Shah Palace, with three different levels including a mosque, mausoleum and a courtyard full of carved stones. A lot of renovation is going on and there were good views over the city and the Caspian Sea.

Baku has a wonderful mix of peoples and cultures, Azeri, Persian, Turkish and Russian. There appears to be quite a lot of wealth and lots of beautiful Baroque buildings, shops and restaurants. Wandering on we needed to find the travel agent who helped us get the visa. The owner Serhan was very friendly and helpful. After inviting us to dinner at his house tomorrow, he walked with us to Fountain Square and we then had tea in the Café Mozart and a small snack of Turkish-style aubergines and yoghurt/garlic/mint at the Ocakbasi restaurant.

30 March: Baku

Not really over the jet lag yet and still tired. We dragged ourselves up after a fitful night. The hotel is too warm and stuffy, and the window cannot be opened. And the air conditioning is rather noisy. Some people are never satisfied!

Our floor lady said it was very cold outside. Emerging from the door, we were practically blown off our feet by a bitterly cold howling gale. 'Baku' apparently means 'City of Winds' and that's certainly true! In the shop windows there's a lot of smart merchandise on sale – shiny pink dresses, red patent shoes with 8-inch heels and 6-inch platform soles, slinky underwear, etc. Some people must have a reasonable standard of living, though Serhan told us the doorman in his office building earns only £3 or so per day.

On the way to Serhan's office, we passed the railway station – another imposing building with crowds of people outside. At his office the doorman indicated that he was not in, so we waited outside on the steps. Five minutes later, he ran into the yard, looking very embarrassed, and said he was sorry his cleaning lady had not turned up today and he couldn't possibly take us to his home in that state! So instead we went to the Oscar restaurant, where he ordered some traditional Azeri food, not on the menu. String with a mixture of salads including a delicious red leaf whose name I can't remember (aniseed-like flavour) we then had a sort of dumpling soup followed by Azeri lasagne – flat squares of pasta with a mincemeat sauce on top. The hours passed really quickly, drinking a lovely red wine and talking.

31 March: Baku

Being a Saturday morning, lots of young Baku-ites are out in the streets in their finery – bright red pashmina, lots of miniskirts and more ludicrously high platform soles. For lunch, we returned again to the Karvansaray, but this time tried the sturgeon kebab – a solid meaty fish.

In Fountain Square is an old Armenian church. Although it is no longer used as a church, it was full of men playing pool. At one end of the square is a high-rise Radisson Hotel above a prestige office block, and below it is McDonalds.

Baku later hosted the Eurovision Song Contest, and from the pictures it looks to have grown into a ritzy city. It's a bit embarrassing admitting one watches the Eurovision Song Contest isn't it! Only to see what Baku looks like now!

1 April: Sheki

By 7:45am we were in a taxi to the bus station, our days of 'luxury' in Baku over. But it has not been such luxury – bright lights on our balcony have kept us awake most nights, and mysterious phone calls have woken us when we have just managed to fall into slumber.

The whole bus station area is filthy, drab and disorganised, with no timetable in sight. Nobody spoke English. Siân's 'O' Level Russian allowed her to decipher Sheki 10:10 on one bus, so we went to sit in it instead of the ice-cold waiting room. Even after it left the bus took almost an hour to leave the outskirts of Baku. Along the Caspian coast were various oil platforms and even one of the natural burning gas fires that first hinted of the oil and gas wealth below before they were drilled. On a sandy beach were some thatched sun umbrellas – an amusing sight!

Soon we turned inland along a bumpy tarmac road. It was a truly grey, damp day – occasionally passengers would get off in the middle of nowhere and disappear into the mists. On and on we went, eventually stopping at a roadside café. At least I suppose it was a café – the toilets were the worst I have seen for 15 years, since China.

The entertainment on the bus varied from an inane video comedy with men and women chasing each other and fainting, to some quite pleasant Indian-sounding music. Towards the end of our journey a smart young boy of perhaps 15 years old asked if we were English, because he had seen our Lonely Planet book! He also invited us to his house, but unfortunately we were just too tired for any detours in the middle of nowhere. (He also mentioned that his home village was Agdam, just outside the boundaries of Nagorno Karabakh, which is now occupied by Armenia, so he and his family were refugees.) The departure of the Soviets from the Caucasus has left a legacy of troubled border areas that soldiers on to this day.

The scenery has been very green since not long after leaving the Baku coastal area and its semi-desert environment. Just like England, except there are more sheep running freely, unlike in our Foot & Mouth devastated country. At the turnoff to the Sheki road stands a huge double brick-and-stone arch – but what is it for?

At the bus station, we were already lost. Soon enough a taxi came by and dropped us at the Sabuxi Hotel. What a dump – no atmosphere, no customers, no electricity, no more to say! Walking up the road in the light drizzle, we found the Kervansaray, a large beautiful building. The rooms were cheaper than expected ($13). It has comfortable beds, a fireplace, toilet, shower and living room. But no hot water – we can live without that for one night!

The main sight of the old town with its cobbled streets is the Khan's Palace. What a beautiful place – wonderful paintings inside on all the walls and ceilings, a precarious carpet-covered staircase and no lights. Outside, the walls were also beautifully patterned in a nice garden with huge old trees. The museums within the fortress walks were closed, and the whole place seemed not to have seen another soul for years. With dark grey skies and falling rain it was time to sample the much-touted 'piti' – a stew of lamb with great lumps of fat floating in it.

The hotel and street in Sheki

2 April: Tbilisi

The so-called direct bus to Tbilisi was slated to leave at 8am. At the bus depot there was no direct bus to Tbilisi, so we were put on the local bus to Qax and told we could get one there. This bus stopped absolutely everywhere and took over hours to cover the short distance.

Disembarking at Qax, of course there was no sign of any bus to Tbilisi except on the timetable. Someone indicated there might be one tomorrow? Realising the only option was to take a private taxi, we went over to use the public toilets – it took three attempts before I could force myself to go in. There a policeman almost arrested us for inadvertently sitting on the Martyr's Memorial while studying Lonely Planet and wondering what on earth to do next.

Disentangling ourselves from the policeman and his requests for 'documenti', we hurriedly went back over to the taxis and cleared off to Balakan. The sun broke through, treating us to a gorgeous view of the Caucasus Mountains. Horses and carts worked in the fields, with the snow-capped peaks beyond. Driving like a loony at times, our driver made it to Balakan in about an hour, despite the potholes. Balakan is a small town with nothing really and certainly we saw no taxis. The bus driver kindly took us on to the border. It was a surprisingly quiet and remote border with no buses, no taxis, no café, no money changer, no nothing.

Although somewhat intimidating, the guards seemed more interested in our guidebooks than money. They asked if we were going to Armenia – the great Satan – then we were so pleased that we hadn't got those visas in advance!

On the Georgian side there were no vehicles in sight. So we walked perhaps 1 km away from the Customs post then sat on a bench in a village to eat our cheese sandwiches made from breakfast. No-one seemed to notice us – a horse and cart passed by, and an old truck parked opposite gave Bob two perfect photo opportunities.

On the Georgian side of the border

We walked about 5km before crossing the wide river and the village square in Lagodekhi. And what should we see but a comfortable Ford Transit minibus, going to Tbilisi. Phew! In Tbilisi a crummy old Russian hotel was eventually found up some dark steep steps and, after a brief rest, we left in search of a restaurant. The Dukani served the local phkali (paste of spinach or beetroot with walnuts and garlic), plus fried sulguni cheese with a delicious bean dish to savour.

3 April: Tbilisi
Again there was drizzly rain, but the setting of the Georgian capital is fabulous. High turreted walls guard the city's hilltop and the whole area below is a mix of traditional wooden houses with just a few ugly Soviet bits. We visited Metekhi Church, dramatically set on a cliff above the river. The tasty local lunch eaten by all the locals is fresh melted cheese oozing from superb flaky pastry. The people have a reputation for music and drinking. There are certainly a lot of wine cellars to be found across the old city. Elsewhere there are many mansions and quaint houses mixed in with

stylish Russian churches and even a blue monastery. Along the main thoroughfare, Rustavelis Avenue, is the famous Opera House, more Russian-style civic structures and some much sterner-looking government buildings

Tbilisi dominated by its old fortress, Georgia

4 April: Yerevan

Waking at 6am, it was still pitch black outside the hotel with no signs of life. There was absolutely no water, neither hot nor cold. Nor was the breakfast room open, so we walked out very disgruntled into the dark, wet, rainy streets. It seemed a long way to the bus station.

Here at least we were able to get fresh hot cheese pastries for breakfast, and one extra for lunch. But no tea, and anyway we dare not drink before these mammoth bus journeys. The bus for Yerevan left by 8.30am but only drove around the corner for a couple of hundred metres, where it stopped for someone to load boxes and bags of something into the luggage compartment. The windows were steamed up and the nearside wiper was missing, so we saw practically nothing of the scenery. Low cloud and mist reduced the visibility to virtually nil.

The road was narrow bumpy and no wider than our lane at home in Colworth. Potholes, deep ones, everywhere, meant our ride was like a slalom course into the unknown. Hardly any traffic passed us in three hours and we had a horrible feeling we were being driven off on some illegal route into Armenia with a busload of contraband!

Hours later a customs shed loomed out of the mist. Here we had to get out of the bus – Don't you speak Russian? Nyet! – only to find that we had not been given the correct paperwork at the Lagodekhi entry border. To extricate ourselves from this problem, $10 each was required. If we had seen a bus going the other way back to Tbilisi, we would probably have taken it!

At the Armenian immigration, they spoke a little English and things went smoothly, if a little slowly for the others on the bus. Back on the road, the countryside changed immediately to treeless rolling hills like Eastern Turkey. Crossing high passes with lots of snow around below the peak of Aragats, the bleak weather and bleak countryside continued until we reached Yerevan. Driving into Yerevan, it all looked very drab and tumbledown.

Exhausted and starving, we changed money at the bus station then took a taxi to the Hotel Erebuni. The nine-storey Soviet block was full of life and a bit like an Iranian hotel. Our room had wonderful views over the derelict buildings and backwaters of Yerevan. Nearby a massive stone arch led into the most grandiose circular Republic Square, with impressive buildings all around.

Central Yerevan, Armenia

Much of the city has wide clean streets, there are quite a few smart clothes shops and everyone seems well dressed. Quite a few people have mobile phones. In our hotel the disco was playing loudly and the beat reverberating through the floor. But the water was boiling hot, the shower had good pressure and it was such luxury after the expensive waterless dump in Tbilisi.

5 April: Yerevan
Our plan was to visit Echmiadzin, one of the holiest places in all Armenia just outside the town. To our amazement and delight, the cloud lifted gradually and we could see quite clearly Mount Ararat and Little Ararat! Arriving in Echmiadzin we were told to get out at the first church. This turned out to be the beautiful St. Hripsime's Church set in a serene and willowy complex. It was, however, not the main cathedral.

Echmiadzin Cathedral has some superb and impressive frescoes but the lower white walls were being repainted and there was scaffolding everywhere. The Treasury and the ruins of the pagan temple underground were impressive. We had tea in a nearby café and they refused to accept any payment. So kind!

Armenia is a mainly Christian country but there are a few mosques, including one impressive blue-tiled version. The Cascade is a concrete monstrosity built in Soviet times. It probably looks better when water is cascading down it; otherwise it's just a huge concrete staircase. Not far from town are the two historic sights. Garni is a Pagan temple from the first century AD, similar in style to the Greek Parthenon. It is perched high above a deep canyon in spectacular mountain country. The cave monastery and church of Geghard is partly dug out of the rock – an incredible monument, dating from the 13th century.

It's been a great day out in Armenia, but there's so much we could have seen if only we'd had more time on our visas. So we returned to Tbilisi satisfied with our brief trip.

Geghard church, Armenia

Mtskheta church, Georgia

7 April: Tbilisi
The historic cathedral of Mtskheta a few miles from Tbilisi is a stirring sight surrounded by vast fortress walls. A wonderful old bell sat in the courtyard. Inside the church were huge arches and beautiful frescoes painted on the walls. It was the Saturday before Palm Sunday with some very special celebrations. Hearing a beautiful male voice choir singing, we assumed the choir was hidden behind the altar screens. But on walking up the other side of the church, we saw an unruly

group of young men dressed in jeans and bomber jackets – the glorious sound was emanating from their lips! They looked like the sort of crowd you might cross the street to avoid in Britain! The nearby sights included a nunnery and another tiny little church in its grounds dating from the 4th century.

Georgian band at Mtskheta

8 April: Akhaltsikhe
We found an old Ford minibus going to Akhaltsikhe. Passing through many pretty country villages, our route took us below rolling hills and rivers glistening in the dappled sunlight. There were good views of the Caucasus Mountains before reaching the town. We had no idea what to see or expect; the main aim had been to see some countryside and the Sapara monastery. The town was extremely rundown and very poor. The old Soviet concrete hotel was a bombed-out wreck of its former glory. The stairway was crumbling, pipes were rusty, there were no locks, no curtains, no water, just a filthy squalid toilet at the end of the corridor with no shower. What happened to the sauna mentioned in the guidebook? We had to eat a small bread roll each and share a tomato. And the one chocolate Lion bar from Tbilisi.

A very friendly old taxi driver took us to Sapara monastery. A dreadful muddy lane led to the church, fortress and some strange beehive-shaped remains. Along the way our driver pointed out lots of things in Russian – we could only look, nod and smile! On arrival our driver came with us carrying lots of empty plastic water bottles, which he filled at a spring there. Was it holy water or just good drinking water? He showed us around as best he could in Russian, explaining that he had been a monk there as a young boy. We marvelled at the quality of the frescoes inside – still covering most of the walls, the colours seem as vivid as they must have been in the 14th century when they were originally painted.

Back in town on our way up the hill, a young man speaking English came running up to invite us to tea in his house. Immediately we were taken into his courtyard, underneath the framework of vines that were being pruned in preparation for the summer. We were offered tea or coffee, and given a lovely book, an English translation of an Armenian poet's work – we look forward to reading it. His father was out working in the garden, but his mother, Ophelia, was very attractive and sat listening to our conversation even though she could not understand a word.

Sapara Monastery, Georgia

Sergei told us he is 23 years old and is studying history at the University of Yerevan in Armenia (previously he studied law for 2 years). His family came from Erzerum (now in Turkey) almost 200 years ago, when they were forced to flee by the Turks. His mother's family were all killed, except for her grandparents (who were presumably young at the time and this must have been in 1915 or so at the time of the Armenian genocide).

Now they are living in Georgia, with Georgian passports, but the Georgian government do not care about their town because they are all Armenians – they frequently have no electricity. And because he does not have an Armenian passport, he is treated as a foreigner in Yerevan and has to pay fees etc as a foreigner. What a terrible situation.

Inside their house was very nice, clean and smart with lots of interesting historic photographs. Our tea was served in beautiful dark blue bone china cups with a 1cm gold design around the top. When I said how lovely they were, he said they were antiques. His mother then brought in an old Armenian vase/lamp stand, which they said had been in the family for 250 years. We didn't dare to touch it!

After two hours of their hospitality, we had to leave, with only a Mars Bar and an English penny coin to offer in return. Out in the street – at least the rain had stopped – we were still unable to detect the slightest hint of food anywhere. So we returned up the cranky staircase to our hotel room. For dinner we consumed the remaining three bread rolls with a minute tin of tuna from Greece – truly a case for emergency rations! Thank goodness we brought extra water!

Putting on our thermal underwear, we settled down to shiver through the most miserable night yet.

9 April: Tbilisi
We left unwashed, unshaven and unfed. The journey back was uneventful; through the mist we passed the gloomy town of Gori, where Stalin was born, an exceedingly miserable-looking polluted industrial city. It explained a lot about the man perhaps. Back at Marjanishville we dashed into the Turkish-style cafeteria for lunch. Food at last! Two plates of delicious beans, two of chips, with herb and onion-flavoured tomato sauce, followed by a spicy walnut cake and of course tea.

10 April: Tbilisi
We visited a couple of travel agents on behalf of Steve Dallyn (from the Exodus days) who was now operating his own wine trips. Fortified by another cup of tea, there was just time to climb up a steep cobbled street to the fortress, where we were rewarded by a great view over the city

For some evening entertainment we were invited to a theatre. It was fun, though we couldn't understand a word. Father Ubu was the name of the play, held in a small room with 40 people watching on three sides of the rectangle. In the floor were various trapdoors which actors disappeared into and sprung out of – a mediaeval couple, army captain, two court jesters (one was the splitting image of Mr. Bean!) a king and queen with quite elaborate costumes.

11 April: Tbilisi Airport
Our flight check-in was at 4am on 12 April, so it was ghastly being at the airport all night. With no heating in the building, it was very cold, and the only places to lie down were on the floor or across metal seats with protruding bars. We tried both these options, but without much success. Before dawn the cleaning lady came round to wipe the floor with a wet mop and disturb our 'slumbers'…

12 April: Chichester
… Somehow the time passed, and eventually we arrived back at Heathrow.

Imagine our surprise the next day when we were in Chichester Post Office and found the chairs there absolutely identical to those in Tbilisi airport! …

Nomads of the Sahara

A collection of short stories from the Sahara.

Mauritania: 1993

The engine spluttered badly and the car slowed quickly; it sank down into the sand and all of us piled out with resignation. Not a man to be ruffled, the driver lifted the bonnet slowly and took a long step back. He reached for a mat and, looking around, decided to arrange his mat in a fairly precise easterly direction. It was time to offer thanks to Allah and pray for a speedy Inshallah – God-willing – repair. We also climbed out, legs aching from the unnatural positions we had been crammed into for the last six blistering hours. There was nothing to do now but walk a very long way off into the desert to answer the call of nature. It was time for a dried-up piece of bread and water.

We had seen one of those late-saver charter fares and were able to take a 3-week option to Gambia. The flight was comfortable, the food was good and the entertainment was great. The atmosphere was distinctly upbeat, with happy holidaymakers anticipating the sunny weather and welcoming smiles of the Gambia. No problems at the arrivals in Gambia; the tour buses were waiting outside. For us of course things now changed tempo radically. Suddenly all the white faces were gone, it was time to don our rucksacks, light as they were, and trudge down the airport exit road to the main road into Banjul. We of course weren't going to fork out for a taxi – we were real explorers. As luck would have it, we were almost immediately offered a ride into Fajara with a local expat in a Land Rover – it had to be, this was Africa.

Our first problem was to see if we could obtain a visa for Mauritania, our objective. This was not possible in London, as they simply do not have an embassy. There was probably an embassy in Banjul. There was, it was hard to find but most co-operative, and only a two-day wait for a visa. They had probably not issued many visas in months.

We crossed the Gambia River by a large ferry and took a share-taxi all the way to Dakar in Senegal. The driver drove at maximum speed; they always do. He never stopped talking to the larger-than-life lady taking up the whole of the front seat. The countryside was increasingly arid, with massive fat stumpy Baobab trees with strange branches sprouting at the tops of the stumps. Dakar, the cosmopolitan capital city of Senegal, was cool and breezy. Modern high-rise buildings line the waterfront and ramshackle old colonial houses hold on to a tenuous existence awaiting demolition. We never seemed to eat on this trip. In Dakar though, it was definitely necessary to try the famous Chicken Yassa, a lemon and onion-flavoured African speciality.

We made a big but typical mistake the next day, by being talked into taking an express bus to St Louis. Of course we were still sitting in the bus station hours later and had hardly left before we could have been in St Louis, had we taken the share-taxi as planned. Share-taxis are normal taxis which depart when enough people have occupied the seats. St Louis is a quaint old French-style town and was the administrative headquarters for the colonists. We stayed at the French Mission, whose pastors would never return to France.

We continued north to the Senegal River and a ferry to Mauritania. The trip nearly ended here. One of our passports fell from the grasp of the zealous official who boarded the ferry at the unloading ramp. It fell as if in slow motion into the Senegal River. By great fortune it was retrieved; the Mauritanian visa did not look a pretty sight. The same officials then claimed the visa to be unreadable and of dubious value! But somehow by nightfall we reached Nouakchott. The passport had dried rapidly in the scorching winds blowing from the Sahara.

The markets of Nouakchott are thronged with some of the most colourful people in Africa. Men in royal blue robes with white turbans move elegantly amongst the crowds. These are the proud Moors – expert traders throughout West Africa and beyond. Of Berber descent, they are the ruling classes of Mauritania, migrants from Morocco and the Saharan oases. We stayed in the Oasis Hotel for one night and then moved to the 'you shouldn't stay there' Hotel Adrar. Cheaper but windowless, it was a seedy house of ill repute.

Pushing on in Mauritania

How many Moors can you fit in a Peugeot 504? This was the question we mused over next morning as we waited for the share-taxi to Atar to fill up. Eleven was the answer, three in the back luggage space, four in the back seat and four in the front seat, including the driver. We got the luggage space seat; it was collapsing under the weight. Still, the locals were exceptionally friendly, passing bread all round for breakfast and lunch, although by that time it was already rock hard. The desert was spectacular through the clouds of dust. The road, once tarmac, deteriorated almost immediately out of sight of the capital. The driver was astonishing, a true master of the sand. The car was magnificent despite the sagging suspension and rattling doors. We crossed mile after mile of deep sandy drifts and only had to get out and push once.

Of course the driver fixed the fuel problem after offering prayers to Allah and we reached the cooler oasis of Atar in ten hours. The town of Atar lies northeast of Nouakchott in the Adrar mountains – a region of canyons and plateau tablelands. Red cliffs and brilliant green date palm groves contrast starkly with the yellowy dust-laden sky. It is the season of the Harmattan – a hot dusty wind that plagues the Sahel regions of West Africa in late winter.

Siân outside Chinguetti mosque, Mauritania

We visited the lost city of Chinguetti, sand dunes seriously marching onwards over the ancient city houses. This is the seventh Holy City of Islam and has a stone mosque at its core. Even the camels looked fed up with the sand. At the Auberge Caravan, a very pleasant watering hole, we stayed in a cool traditional-style room, sleeping on mats on the floor. The hospitable owner treated us to superb carrot and lamb couscous by the light of a hurricane lamp. Sadly this was as far as we could get, public transport meant one solitary crumbling Land Rover taxi leaving perhaps today or perhaps not. We waited in Chinguetti for a whole day before finding the Land Rover. It was no place to be thinking about our very fixed, non-transferable return charter flight to London.

In Atar we met a young Belgian with high hopes of training some local students in the rudiments of mechanics, but his task seemed doomed to failure in the lethargic heat, overpowering dust storms and overwhelming bureaucracy. We managed to fly back from Atar to Nouakchott; the flight, often cancelled due to sandstorms and other reasons, ran on this occasion. From Nouakchott the journey back to Banjul was relatively simple; we just sat quivering in the rear of various share-taxis driven at top speed. On reflection it would be fair to say that the driving standards were quite good but, as a mechanic said, 'we keep them going' but 'we don't go in them'.

The Ténéré of Niger: October 2000

The dark tall man climbed down gracefully from his mount and hobbled towards us. The sand was firm and the vision before us quite stunning. The man begged us for some medicines; his unruly animal had given him a very nasty blow and the wounds were ugly, black and swollen. With over 400km to the next pharmacy, his plight was in the balance.

With some antibiotics, this man will hopefully recover. In most parts of the world, a wound would never become so bad, but here in the desolate dunes of the Grand Erg de Bilma, with only camels and the stars for companions, things are very different. Niger is no paradise, time has stood still; blue-robed nomads scrape a living between wells and oases of date palms.

> **Tourism develops in Niger**
> Niger has one of the lowest standards of living but, despite such poverty, the vibrancy of the people is overwhelming. Emerging from droughts and a long, destructive insurrection in the Tuareg regions of the north, Niger is now hopefully on the brink of a new era of stability and development. The fabulous desert landscapes are once more accessible to those in search of adventure, stark surreal landscapes, colourful welcoming people and moments of reflection and solitude. But for how long?
>
> **2015**: Sadly it did not last long and Niger is not currently safe to visit... but hope springs eternal!

Agadez, the only city of the north, has remained true to its heritage, barely changed in twenty years. The mud minaret of the mosque still dominates the main square. Old mud houses keep their inhabitants cooler than any modern concrete, retaining their aesthetic functional beauty. The colourful markets are beyond description. At the Hotel de l'Aïr, a former Sultan's palace, tea is still served on the roof by a smiling waiter. His brown-stained jacket still fits badly.

Out in the desert on the sandy tracks are hundreds of camels, some trains over three hundred strong, still carrying salt from the remote salt-mining oases of Fachi and Bilma. The vehicles make light of the dunes and travel at over 50kph along stunning dune corridors. How on earth do they know which dune to cross and which to avoid? Only the camel footprints and abandoned night camps give occasional clues.

Dirkou is a strange place, lively and mysterious. All of West Africa has arrived here, mostly travelling on top of huge old, overloaded six-wheeled Mercedes trucks. From Ghana, Nigeria, Chad, Mali and Timbuktu, they all seek their fortune in Libya as guest workers. Some are desperate to go home; others seem to be in limbo in Dirkou, running temporary shops and bars, fortune-seekers waylaid in the desert en route to their homes. Smugglers abound; the market is a great melting pot of humanity. What will be its fate in fifteen years' time?

(Of course we now know in 2015 that many of those migrants from West Africa will be heading even further north from violence-riven Libya into Europe. Is this the result of more misguided action by the people who rely on their own perspectives to run the world? One size fits all!)

In a quiet spot at the end of the oasis by a well sits an old man. Barrels of fuel litter his front porch – if such words could describe his doorway. He regales us with tales of World War Two, of Tobruk and General Montgomery and Wavell; he speaks good English. He offers us the finest bread we've ever tasted (one-week-old baguettes have been our staple breakfast for days). Our gentleman is from Libya and his fame has spread throughout the Sahara; he is the Libyan Jerome.

Happy truckers in Niger

Camel caravans en route to Fashi and Bilma, Niger

We pass the conical rock of Zoumri and before us lies the shimmering ghost-like settlement of Seguedine. Standing in isolation before the village is the new clinic; there are electric sockets in every room, the fans are installed, the light bulbs are clean and the new medical students are overwhelmingly enthusiastic about the new project. But there's no electricity. They desperately need simple medicines. We try our best. Perhaps one day electricity will arrive to turn the cooling fans.

Camping en route in Niger

Camping near Seguedine

From the Col de Sara, an exciting area of wind-eroded blocks, the route follows the base of the Djado plateau. Here are stunning towers, dark brooding rock massifs with organ pipe-like structures as guardians. Deep in the fabulous canyons is simple but hidden rock art. Golden sunsets deflect the light into strange ghostly patterns and the eerie silence is almost unbelievable. To camp here is magical; as the moon rises, the rocks are caught in silhouettes of fairytale proportions. Cool breezes refresh the spirits and the stars come ever closer.

Below the Djado plateau

The ancient fortified mud citadel of Djado is a ghost town almost hidden by date palms. Its tiny, narrow streets are barely navigable between the crumbling mud walls. Concealed in this maze is an incredible Coptic chapel, its cross hardly discernable, sculpted into the mud wall fourteen centuries ago. Mosquitoes are said to have driven out the inhabitants, but sporadic rains sometimes bring seasonal pools to life. Djado is a dying edifice, with long-forgotten romantic links to the great caravans of the Sahara.

Djado plateau outcrops, Niger

Djado citadel, Niger

The Ténéré & Crabe d'Arakao, Niger

After 200km of nothing but undulating hard-packed sand, we still have another 200km of hard-packed sand to go. The horizon is lost somewhere in a trance. Nothing disturbs the skyline. This is the Ténéré, an unbelievable expanse of nothing. No road signs here, yet after over 400km of undisturbed sand our Tuareg guides and drivers have arrived at an exact spot, the Crabe d'Arakao. Here dune is piled upon dune, captured within a forbidding, encircling black jagged rocky wall. This is an amazing geographical feature adjacent to the volcanic Aïr mountain range.

Collecting water at a well in the Aïr Mountains

Villages and isolated settlements of Tuareg herders are to be found among the cones and peaks of the Aïr Mountains. These curious, hardy people run to greet any intruders, eager to sell simple but exquisite trinkets and the famous silver Agadez crosses. At the lush oasis of Timia we find food in abundance in the date palms and orange groves below stark cliffs. We are back into the Sahel thorn and acacia scrub. Our sleep is disturbed; our mattresses are blocking the path of curious beetles pushing balls of dung.

All too soon our convoy heads back into Agadez. We've hardly seen soap and water for what seems an age. Our stained clothing is much worse than the tea waiter's, but we soon relish his tea! Agadez is the end of the desert road; the end of a trip into the past. Perhaps this is paradise after all.

Agadez, a fabulous desert town, Niger

Algeria, the Tassili Plateau at last: April 2001

> *How can we be out of soap? Judging by appearances, we have not been contaminated by soap for weeks.*
> **Stones of Silence, George Schaller**

Ever since an aborted trip to Algeria via Tunisia and then finally in 1993 via Morocco to Algeria, both in our Land Rover, we had been hoping to return to the country to visit the Tassili Plateau.

In 1991–2 free elections elected an Islamist-leaning government. This was perhaps the first experiment with democracy in the Muslim world. In the event the old guard didn't like the way the Islamist party was heading and the army took power. This naturally brought a wave of terror attacks that seemed to go on forever. The Algerian Sahara seemed off-limits for an age.

In fact it was a surprising turn of events when groups started going back to the country. We waited no longer and set off, even in the heat of early summer; the chance to visit might blow away again at any time.

27 April: outside Djanet

Yoghurt, banana and Bakewell Tart in our room were followed by croissants and coffee/tea after we had checked in for our flight – practically the only Europeans on the plane from Paris, except for a group of 5 young French people with boots on.

The flight to Algiers was uneventful. It was amusing to read in the Times and Telegraph that the BBC is saving £20,000 by cutting out croissants, chauffeured cars and management consultants, but even more amusing to see it on page 2 of the Algerian daily paper!

With the help of Algiers airport official reps we whizzed through immigration formalities, out of the international terminal and along to domestic, where our bags were again checked in hurriedly. With only 15 minutes to spare, we were in the departure lounge for the Djanet flight.

Flying first to Ouargla, we crossed a small mountain range to leave the green surroundings of Algiers and find the dry desert beneath us. At Ouargla many people got off, but a surprising number of people also boarded the plane, with baggage being personally identified as in Algiers.

The rest of the flight was quite hazy. Soon we could make out the landing strip at Djanet airport and flew south past it before turning and coming back in to land. Inside the terminal a quietly spoken man came up to us and asked Bob if he was Mr Gibbons. Med (Mohamed of course) was there to meet us; what a relief.

In his open-backed Land Cruiser we drove the 30km into town – no wonder we hadn't seen it from the air – and went into a handicraft shop run by the brother of Lotfi Brihmat who works at the Algerian Embassy in London. Here we had tea and I bought a lovely purply-pink chèche. At least I tried to buy it, but they wouldn't let me pay.

After this we were off into the desert to camp, (after filling in a park entry permit form and stopping to purchase essentials such as cigarettes and orange drinks!). Med decided we would camp right under a stunning rocky tower by the side of the 'road'. Weird and wonderful rock shapes surrounded us in the soft golden light of the late afternoon – it's hard to believe we were in Paris this morning!

Just as darkness was falling, the donkeys arrived. Dinner was barley soup, stew and tinned fruit salad. Then we crept into the nylon dome tent that had been erected for us for a sleepless night, with the wind every few seconds flapping the flysheet against the inner and threatening to blow it away entirely!

28 April: Tamrit

It was getting light at 5:15am and we could hear movement outside, so it was time to get up. Breakfast was bread, jam, honey and margarine, just like in Niger, but this time we were forewarned and had brought our own muesli!

The crew below a typical outcrop in the Tassili

The climb upwards for 1½ hours to the first level of the plateau was achieved, below huge rock towers on the left. At the plateau we stopped for oranges and gazed around in amazement. It was then flat for a while along a narrow canyon, green in places. The donkeys turned left before we did, to take a more gentle route up – so what lay ahead for us?

We soon found out, when we began to climb up an enormous rock face/boulder field. After a slight descent we rejoined the donkeys' route and walked along another flattish valley to find our lunch place under a rocky overhang. Phew. The heat is really too much and millions of flies are following us everywhere (Med told us later that the flies in the Sahara are clean, *mouches propres*, and indeed we did not get ill so perhaps he is right!). Strangely though they seem to be more attracted to the men (including Bob) than me – is it their smell or simply natural pheromones?

Three hours later it was time to drag ourselves up for the last ascent – another hour up an almost vertical cliff, a bit like a via ferrata in the Dolomites, but without the cables and not so steep. Truly fantastic scenery, but it is hard work. (It would be much easier in winter, though.)

Suddenly, we were on top of the plateau. Stretching for miles ahead, an almost flat rocky plain as far as the eye could see. No shade at all. The path was clear across the plain, and we walked with our cook, who said there was much more to see off to the left in the jumble of rocky pillars which now appeared. There are so many weird and wonderful shapes here; it's hard to know where to look!

Soon we were in the midst of yet more huge blocks of rock, sedimentary layers, different shades of red, white and ochre. And on many overhanging surfaces appeared the first of several rock paintings we would see. Below many of the overhangs there was a deep depression in the rock, where the pigments were ground and mixed from other colourful stones. Pictures of cows – some spotty Friesians, some plain – were the most common. Also we could make out people standing or squatting around their fires, gazelles, mouflon (ibex) etc. Some are still incredibly detailed after up to 5000 years; others have been removed by unscrupulous tourists and explorers.

Nearby, beneath huge towers of rock, we set up camp, the donkeys were unloaded and we sat and had tea. No more tents! Just as we were leaving to find a place to lay out our mattresses for the night, the guardian of the site came over for a chat – it's a lonely life on the plateau for months on end!

29 April: Sefar
Again we were awake at 5:15am but this time we had had a fantastically good night's sleep in our Hotel des Milles Étoiles. For breakfast we had a surprise – omelettes – which was just as well, since we had six hours to go (though we didn't know that till lunchtime!)

Tassili towers, Algeria

Then we were off into a magical world of fairy towers, searching for animals under rocks (a sort of lizard), marmots, fennec etc. – all sorts of tracks in the sand. Every so often we would pass a beautiful flower growing out of nowhere. In fact it's practically impossible to find words to describe the scenery here – so vast, so tall, so jagged, so colourful...

So I'll digress to a story told by Med, our guide and director of Maha Tours ...

> **The Varen**
> Once upon a time the Chief of the Tuaregs was on a journey with his camel. While resting on the ground, he was bitten by a poisonous snake. But the Varen, a prehistoric type of lizard, appeared. Killing the snake, he then sucked out the venom from the Chief's leg.
>
> Somehow the Chief found the energy to climb up on to his camel's back, but he had no strength left to make the camel move. So the faithful Varen joined him on the camel's back and, by whipping the camel with his long spiky tail, made him go as quickly as possible back to the village, where the Chief could rest and recover...
>
> And ever since the Varen (which still exists) has been sacred to the Tuaregs...

Along the way we saw lots of cave paintings, and rested under the shade of the towers. Gathering bits of wood, the cook made a small fire to brew up a pot of water for the scrumptious mint and tilleul tea, which even Bob is growing to like!

Eventually leaving the towers behind, we had a huge open plain to cross with no hope of shelter, and by now it was truly the midday sun. The slightest overhang would make our guide Zin Ahmed (aged 50!) dive underneath and rest after making sure we wanted to as well! At one such point he gave us a crocheted telephone cover and bracelet, showing us also a woven belt and Tuareg leather purse which his friend wanted to exchange for a sleeping bag, but which did not appeal to us! But we had to continue into the blistering heat of the midday sun again until we reached the edge of Sefar – a huge massif of rocks criss-crossed by deep rifts, and of course overhangs.

It was under one of these huge overhangs that we found our kitchen and campsite at about 1pm. Time to slump and wait for our salad lunch – delicious cucumber, olives, beetroot, tuna fish, tomatoes etc. and fresh oranges. Showing the family photos and the wedding photos caused a lot of entertainment all round.

More outcrops in the Tassili

At about 4:30pm we left again on a voyage of exploration into Black Sefar, so-called because of the darker rock. In this maze of small canyons, lush green plants with huge vivid pink flowers somehow found the means to bloom. And the variety of paintings was incredible. After a late dinner we lay out our mattresses under a nearby overhang, which was not such a good idea as a rather hot night lay ahead. A fennec ran away just a few feet in front of us on our last nocturnal excursion!

30 April: Donkey Camp

After breakfast we explored the area around the campsite but dared not go too far for fear of getting lost in the maze. Zin Ahmed took us on a tour of White Sefar where, among others, we would see the huge painting of the Great God of Sefar.

Tassili art, Sefar

At this point it was time for the fancy dress party – Med brought out his party clothes and Bob was dressed by our cook to look like a real Tuareg. With his dark skin and brown eyes, plus unshaven moustache, he really looked the part. We took a slightly different route back through the towers of Tamrit. The scenery is truly unbelievable – not just the prehistoric art but also nature's own art.

The gang in Sefar, Tassili

For some light relief from the heat, we had some jokes from Med:

Light relief in the desert

1. How do you get an elephant into a fridge in 3 moves?

2. How do you get a giraffe into a fridge in 4 moves?

Some time later, but not late enough, came the last joke in the series…The king of the animals had a party and everyone was invited, but one animal did not come.

3. Which animal was it, and why?

See the last paragraph for the answers.

Tonight it was my turn to make the taguela, a type of bread cooked in the sand. First of all they make a fire with wood. Meanwhile the semoule (a flour made from corn, I think; it's yellow and looks like polenta) is mixed with water and a bit of salt, kneaded into a smooth dough, then pressed into a circular shape and tossed in a little of the dry semoule. When the fire has turned into charcoal, a hole is made in the hot sand, then the bread mix is placed in it and covered over with more hot sand and charcoal. For about an hour it is left alone, then tapped with something to make sure it is hard enough to come out! It sounds like solid rock! Cut or rather torn into small pieces, it is then mixed with a vegetable stew, which has been cooking on another part of the fire. We were able to taste a tiny piece of the taguela before it was cut up and it was just about edible when fresh from the fire!

209

1 May: Djanet

As usual we were awake early but so were the others and we had to wriggle more awkwardly than usual to get dressed inside the sleeping bags!

Going down we took a different route wherever possible to vary the scenery. After a welcome drink in the shade of a small building, it was time to bid farewell to the donkey men, the cook and our local guide. Saying goodbye to our cook, we asked him to write his address on a card so we could send him a copy of the photo, but surprisingly, although he has eight children and speaks five different languages, including French, Arabic and three African languages, he cannot read and write.

Now it was time to start the next phase of our adventure... Since Med and his team are returning to Tamanrasset after our trip, he had suggested that we might go with them to the Hoggar Mountains, so our flight was changed.

For lunch we drove out almost back to the airport to visit the 'garden' of Nadir Brihmat, the man who runs the artisan's shop in Djanet, whose brother Lotfi works at the embassy in London. We were invited to dinner tonight. The garden is actually a huge complex of date palms, tomatoes, lettuce, aubergines, oranges, etc., owned by the extended Brihmat family, and they are also building some en-suite rooms for guests.

Before sunset we drove into the dunes to see the engraving of La Vache Qui Pleure (The Crying Cow). There are two engraved cows on a rock piton surrounded by dunes – the first engraving we've seen here. It's more than 5000 years old, but so clear it looks like it was done yesterday. Our next stop on the way back to Djanet was the Hotel Ténéré Village – a lovely place built eight years ago but sadly hardly used yet. Back at our campsite in the Brihmat nursery we found lights strung across the courtyard, carpets, tables and thick cushions on the floor, and Coke and mineral water on the tables. After a noodle soup we had a fresh garden salad plus about five grilled lamb chops each; deliciously tender meat and a great atmosphere.

As we were preparing to leave, Med told us to remove our shoes again as he had a surprise for us. Then a man appeared with a tent pole, which had four holes drilled in it, and proceeded to play magical tunes on it as if it were a flute! Our cook also played it and got two notes out of it at the same time; a lower continuous sound plus a higher melody! Wonderfully dreamlike music! Something I must try at home!

2 May: in the dunes

What a shame to get up so early in such a comfortable place! But we had a lot to do...

So we were off earlier than expected. Slightly south of Djanet, En route we stopped to walk into the canyon and guelta of Essendilene. We retraced our steps back to some huts where a Tuareg family live and our lunch spot under the trees. For 3 hours we sat under the trees, moving only as the sun's shadows changed position, eating incredibly juicy oranges to quench our thirst and keep our blood sugar levels up! Med told us that he had given a pot of honey to our guide's father, who is the principal guide here – his wife had had a nervous breakdown, which he cured with honey and a special brew of local herbs.

By 4pm it was cool enough to move on. Around sunset an outcrop called the Les Seins de la Négresse was passed. We camped in an area of small dunes near some boulder outcrops. Of course Bob had to go and explore the rocks, even as the sun was beginning to fall, so decided to wait behind a bush and write my diary. As the sun was fading fast, I was relieved to see a blue spot coming towards us across the sand; otherwise we would have to send out a search party to find the errant explorer! Tonight it was Bob's turn to make the taguela, as he discovered to his horror when he returned (only because of the sticky fingers it entailed!).

Sitting round the fire, we could just make out the lights of Fort Gardel in the distance, and as the night fell they became brighter. I wonder if they could see our little fire glimmering in the distance? By moonlight, we made our way downhill to a spot slightly hidden by a low bush, where we would lay down our mattresses and sleep through the cool desert night...

3 May: near Ideles

It was to be a long drive today, as we had 500km or so to go across the open countryside to reach the mountains of the Hoggar and Atakor. Driving initially through sandy areas, we passed a still-occupied fort originally built by the French, where our vehicle and driver's papers were checked but we had no hassle. It must be a hardship posting to be sent out here for months, but the peace and solitude must be a bonus for some of these nomadic people. With no other human beings to be seen or heard anywhere, our only company en route was the odd solitary bird of prey and gazelle, beautiful delicate creatures who somehow survive in this incredibly harsh environment.

Typical outcrops en route to Tamanrasset, Algeria

At one flat spot we stopped, I thought for a cigarette break, but when I returned from an excursion into the rocks I found that we did in fact have a puncture, our first in this rough rocky terrain. The wheel was quickly changed and we continued on to the tiny settlement of Ideles, where luckily they still had supplies of fuel. But there was 'no vulcanisateur here!' to fix the puncture.

Camped in the rocks out of the village it took the four men, including Bob, over three hours to remove the tyre from the wheel, it was stuck on so tightly. Dinner was rather late because of this and we did not get to bed until well after 10pm.

4 May: Assekrem

After Hirafok the route climbed up and up and up into the Hoggar Mountains. What a road! At some places we had to get out and walk, the road was so steep as well as impossibly rough. En route we met a solitary Frenchman on a camel, with two other camels and a guide. He was doing a three-week méharée i.e. camel trek through the mountains. Wild donkeys en route apparently date from the time of Père Foucauld, the French priest who founded a mission here but was subsequently accidentally murdered. All around massive volcanic plugs jut out of the sandy granite, with shapes that defy the imagination. A few small birds with attractively coloured feathers fly around; there are also gerboise, small rodents, which supply a source of food to the larger birds such as crows and eagles.

Assekrem approaches, Hoggar, Algeria

In the middle of this dry rocky massif, we suddenly turned down into a hidden guelta where the massive plants with their beautiful pink flowers really flourished. Still more weird and wonderful shaped rock outcrops lay ahead, too amazing to describe in words. By 3pm we had reached Assekrem refuge. The holy chapel, the hermitage of Père Foucauld lay just a little further up the hill. The altitude is around 2800m so it's not surprising we felt a bit breathless. A French priest appeared and invited us to have tea with him after we had seen inside the chapel. It was the most delicious cup of Twinings Earl Grey tea I have ever tasted.

Before dinner, we decided to climb up the hill on the other side of the col. The view was wonderful, the soft evening light illuminated the peaks opposite; through the clouds, shafts of different shades of light created shadows and perspectives on the layers of mountains in front of us. We returned to the refuge for our dinner, another meal of soup and taguela, in the refuge. A French group appeared late having completed a two-week trek in the Tassili du Hoggar further south of Tamanrasset.

5 May: Tamanrasset
Passing through the mountains, Bob felt very nostalgic for the times he had been here before, in 1975, 1976 and 1978.

Quite close to Tamanrasset, we stopped again with a puncture. Of course sometime later some cars stopped by our vehicle and offered to help. The driver of one vehicle offered us his flashy expensive spare wheel. He was an Algerian tourist from Annaba on the north coast, in his own vehicle, and was travelling with a local guide in another vehicle. Being a diver, he also had an adapted compressed-air cylinder, which we used to increase the pressure on the other tyres! They were going to Assekrem for the night and would return to Tamanrasset tomorrow, so we could leave the spare wheel there for them to collect later.

Soon after this timely rescue, we were sitting in a restaurant in Tamanrasset eating camel steak, chips and rice, followed by scrumptious oranges and preceded by a delicious homemade vegetable soup made with locally-grown fresh vegetables. The waiter/owner has a brother who is married to an English woman, and he laughed as he told us she was a British Indian Muslim who wears the chador.

Tamanrasset has grown into a town-sized settlement in the 23 years since Bob's last visit. It was quite a shock. Med took us first to his brother-in-law's house for coffee and cake. His wife's brother is fair-skinned with blue eyes, and comes from the far east of Niger near the Chad border. They were brought up in Zinder on the Niger/Nigeria border. Apparently their father was a French militaire. There is a quite extraordinary mix of people here in Tamanrasset – Med told us there are people from 44 countries here!

6 May: Chichester

Despite a planned departure at 6:30am, the check-in was delayed for no apparent reason and then, just as we were about to board, they informed us that the flight would not stop at Djanet but go straight to Algiers. Fortunately this did not affect us now, but we wondered what would have happened if we had not changed our flight – would we be stuck in Djanet for an extra day?

The flight to Algiers was magnificent – we could pick out sharp-edged rocky plateaux below, then a vast extent of dune corridors which seemed to go on for ever. At Algiers there were a lot of security checks, but nothing intimidating. The northern part of the country is still under siege from Islamic militants and they periodically murder people for reasons known only to themselves.

And then, before we knew it, we were back in Heathrow with Beryl and Tony waiting to meet us. It seems very strange this time to have come so far during the same day, last night in the middle of the Sahara and tonight in the middle of the south of England.

Our safety never seemed in doubt, so we have written to the Foreign Office in London to change their travel advice so that British people, like the French, can experience the Sahara and its wonderful people for themselves.

Postscript: Of course things got worse a few years later, as we record in the chapter, Libreville or bust.

Answers to jokes

Answer 1: Open the door, put the elephant in, close the door.

Answer 2: Nothing to do with bending its neck…
Open the door, take the elephant out, put the giraffe in, close the door.

Answer 3: The giraffe, because he was stuck in the fridge!

Chad: 2002

It had been at times frustrating trying to organise this trip. The trips only go when there are enough clients for two vehicles. We had hoped to get to the Tibesti, and still do, but there are other treasures in Chad to discover.

For as long as anyone can remember, all of northern Chad has been one of those 'only essential travel' places that are forbidden by many western governments. This means no valid insurance for those who do venture into such territory. And who can say for sure how safe any such place is; but if we wait for the travel ban to be lifted we'll never go or we'll be too old.

19 February: N'djamena

Flying down over central France, we could see the Alps in the distance, poking up through the equally white clouds.

We arrived in N'djamena at 10:10pm. It took a long time to get out of the airport. Behind the glass panel we could see a man holding a pink placard for Mr and Mrs Siân, but it was over half an hour before we could attract his attention. Then Hassan, from the agency, came and took our passports away, taking us out past immigration to collect our bags and wait in their office at the airport. It seems we are now five for the tour, as the Italian has not turned up. The airport is quite smartly painted and seems fairly well organised for an African one!

By midnight we had reached our hotel, still without passports. The Hotel Sahel is very pleasant – a central courtyard surrounded by single storey large rooms with hot water and CNN, and wow! air conditioning and fan.

20 February: north of Massakory
Up at 8am, to meet Hassan after our last English sandwich. Moussa, the boss, is in Mecca on pilgrimage, but his Swiss wife (from Basel) was there, so we were able to discuss various topics, such as the state of tourism here. Apparently the French consul has refused to allow embassy staff to go into the desert, but has gone there himself on holiday! And according to the Rajasthani Indian who manages our hotel, the total number of tourists visiting Chad last year was only 216.

We spent some time sitting in the car outside the Direction Générale de Sécurité Nationale, waiting for our visas to be processed, but yet more paperwork was required. So we returned to the office for Hassan to sort it out.

We walked in the heat to the Novotel, where Bob had camped back in 1978. It's a shame you're not allowed to take photographs anywhere in N'djamena – the streets are wide boulevards, lined with trees, and very quiet. Some are tarmac but most are hard sandy earth – the sand of course pervades every roadside and courtyard. Along the verges are some small stalls selling fuel in old alcohol/Pernod bottles, and others selling the ubiquitous baguettes, bananas and avocadoes. Armed with four bananas to supplement our remaining biscuits, it's back to the room for a last shower. Before leaving, we discovered that the Rajasthani knows Pram and his daughter in Jaipur and worked in the Bissau Palace Hotel shop for a while.

Just before 2pm, our car arrived, and soon after Hassan appeared on his motorbike, followed by the other car with its three passengers.

The tarmac road to Djermaya passes through flat, open scrubby land with some small villages. Stopping for cheese just 20km from N'djamena, we were able to briefly introduce ourselves to the other three passengers; Marie-France, Thierry and Christophe. At Massanguet we left the tarmac to head north. The villages have lots of pretty houses, like mini-fortresses of mud, with added-on round roofed 'conservatories' made of golden-coloured straw. Long-horned cattle are in abundance; there are lots of goats.

Just after Massakory the sun slipped below the horizon, and we slipped off the piste to find a campsite. We have dome tents, a table with tablecloth and armchairs to relax in. For an aperitif we had orange juice with peanuts, dates and olives, and for dinner couscous with a vegetable and meat stew. Delicious. And dessert was fresh mango, sliced in half off the stone and then cut through into blocks so the skin would fold and you could bite off the cubes. This trip could get to be quite fattening at this rate.

21 February: 'Melons'
We discovered that Thierry and Christophe are both teachers at the French Lycée, and Marie-France is visiting her daughter, who lives and works in N'djamena.

Breakfast was wonderful – bread and strawberry jam, an unlimited supply of omelettes and even a packet of Swiss Co-op Muesli. A couple of young boys appeared from nowhere to take our empty plastic water bottles.

It's a long morning's drive along the Bahr El Ghazal with palm trees growing in the dried up ancient river course. At Moussoro our travel permits were handed in, and we filled up with diesel from the local shop. No funnels here – it was done by siphoning from a jerrycan held high above the man's shoulder. It was a fascinating little town, with lots of people riding camels. And a donkey train passed by, with a baby goat riding on the back of one of the donkeys. Such a shame we are not permitted to take photos... but we did sneak one on the way out!

Finding shade for lunch was not easy. The afternoon's drive was long, through nondescript empty countryside, and we arrived at camp just too late to capture the setting sun on film. Our campsite was strewn with melon-like spheres, and dinner was avocado and boiled egg salad with potatoes peeled by Bob.

Extraordinary Times in Chad

22 February: Faya
We found out that Thierry is an English teacher, hence his reticence at first to tell us his subject! Along with breakfast, Nour, our driver, brought over the GPS, so Siân has a new computer to play with! The day's drive was mostly hot and uneventful. For lunch we stopped on a wide-open flat sandy plateau, hiding in the shade between the two vehicles.

Around late afternoon an enormously overloaded Libyan truck was stuck in the sand with 30 passengers languishing in the baking sun. At least half of them were women, babies and young children. It appeared they had been stuck there for two days already. Our drivers tried to help them to get started, but to no avail. Their batteries were dead and the alternator was not working. So all we could do was take one of their men with us to Faya, where he would try to get help. It looked a terrible situation, but they must have had food and water, as they didn't ask for any. They were all very friendly and in amazingly good spirits, especially the women and children, who were happy to talk to us and shake hands, and even have their photos taken.

As the sun faded away the oasis of Faya was a welcome sight. There are green and lush palmeries, a military airport, and a beautiful mud town. Driving through the narrow picturesque streets, everyone waved to us and the driver joked to us that we were going to a hotel. But it was true! The Hotel Restaurant Emi Koussi was to be our night stop. Not exactly luxurious, but we could pitch our tent in the secluded courtyard and have a cold shower from a bucket.

23 February: near Ounianga Kebir
A lie-in today – breakfast was due after 7am, so we took the opportunity to go exploring the empty sandy streets and mud buildings of Faya. It is a lovely town. Out beyond the mysterious alleys are some palm trees and low dunes on the northern edge of town. People were beginning to stir from their cosy cool mud houses by now.

A young woman approached us. She was 25 years old and had been forced to marry at 15. She says she has three children aged 10, 8 and 6. They live with their father in N'djamena, as she is now divorced. But she flies to N'djamena with the French military aircraft every two months or so to see them. She had to leave school after class 6 but seems very intelligent, and is now working for the authorities for £40 per month.

Faya mosque, Chad

Because the route close to the Tibesti north is forbidden by the military, we headed southeast at first and then northeast, stopping for lunch between two wonderful rock formations, surrounded by golden sands. Continuing on across the plains, we pass the forbidding plains of Wadi Doum. Some terrible battles took place here during the Libyan/Chad war. We camped just off the side of the piste. The wind was foul making pitching the tent dreadful. A constant stream of sand whistles through all the zips and gaps in the tent. We wonder if we'll ever get to sleep outside on this trip?

Ounianga Kebir lakes

24 February: near Ounianga Serir
What an exciting day, with the most stunningly unbelievable scenery imaginable! We feasted on deep blue and green lakes, golden sand, vivid green foliage and a deep blue sky. The sensational and beautiful lake of Ounianga Kebir is surprisingly vast. A deep blue stripe above the unfolding dunes is just the first glimpse. The town is a shabby mud settlement with flies and windblown kids in tatty clothing. We fill up with fuel and water and are able to explore on our own.

Various Libyan trucks were filling up with passengers, being already overloaded with supplies. We managed to sneak a few surreptitious photographs while no one was looking.

From here it should have been only 56km to Ounianga Serir, but we got lost! Ismael, our lead driver, came running back asking Siân for the GPS reading. (I had been looking at it anyway and wondering why we were going NE instead of SE, but with yesterday's route being completely different from the map, I had not been unduly worried.)

Cutting across unmarked territory to rejoin the correct route, was slightly concerning in case there were leftover mines. An unfortunate young man from Cameroon, whom we had just met in Ounianga Kebir, had been injured by such a device. But we survived without incident – passing through some of the most incredible desert scenery imaginable. Enormous wide pristine sandy corridors along which we sailed – it almost felt like sailing – between vast rocky outcrops of black and various shades of red and gold. Fantastic!

Around Ounianga Serir are more, smaller picturesque blue lakes surrounded by vivid green palms and golden sands. Soon the atmosphere of tranquillity was interrupted. Children previously silhouetted against the sky in the distance came closer and closer. They were trying to sell us rocks and some rather nondescript jewellery but one had a commemorative crown issued during the wedding of Prince Charles and Lady Di!

Ounianga Serir lakes

Above the lakes high in the dunes the camp was exceptionally beautiful with red outcrops and golden sand. A gorgeous sunset was followed by the starriest night imaginable. For three hours lay outside, but were forced to retreat into our tent for sleep by the all-pervading sand, sucking away the humidity to turn us into Vieilles Prunes, like the Swiss liqueur!

25 February: in the dunes
As usual, we had just fallen asleep when it was time to get up.

At Tegguidé was another lake, yellowy brown in colour and sulphurous. The area is not inhabited. Date palms grow in abundance around its shores, and a small spring provides fresh water to the people who come to harvest them. But the spring was surrounded by swarms of mosquitoes – not a pleasant place for permanent habitation.

From here banks of dunes often blocked the way and occasionally the cars got stuck in the sand. We reached the apparently idyllic village of Demi, below two impressive peaks. The Tubu women were collecting salt from the earth. Quite literally, they would dig into the red sand with a spade and drag up glistening crystals of salt. This mix of salt and earth was then put into a circular dish and shaken in a rotating manner, like gold-panning. The salt was then lifted off the top and stored in another bowl. When asked politely and discreetly if we could take photographs, their manner changed and they became extremely rude and unpleasant. A gang of kids had also accumulated and they were so rude that we were all forced to leave. Even taking pictures of the ground was not permitted. It was a great pity as children are normally instinctively friendly.

Tegguidé Lake & saltpans

Bob had already been watching all this from a nearby hill topped with palms and had already taken some panoramic pictures of the amazing landscapes. Our party headed on towards the Depression du Mourdi, a vast cauldron of shifting sands, boulder littered sandy plains and some small shimmering white salt pans.

In the middle of nowhere, a sandy area full of dunes, we came across an old Toyota car, fully loaded with sacks of provisions, and a guerba (goatskin containing water) on each side – the first guerbas we have seen out here, though they are common in Niger and Algeria. There was no sign of the occupants – we can only hope they have gone off in some other vehicle to get help. Soon after this we pulled off below a large dune and set up camp. A howling wind blew sand everywhere again. The two chickens, who had travelled on the roof of our car in a mineral water carton, now met their sorry end to provide us with a chewy meal. That night we ended up covered with sand as usual. Cloud was building up in the sky.

Panoramic view at Demi oasis, Chad

26 February: Fada
After chewing the last of the N'djamena bread, we were off across the dunes, constantly getting stuck and losing the piste. After discussions with the two drivers and cook, we determined that we were heading east along the Depression du Mourdi to approach Fada from the north. But after driving east for about 30km, they met another vehicle and decided to go back, meaning that we had wasted about 4 hours. And to cap it all, the skies were black and we even had some drops of rain, heavy with sand as they fell on to the windows. Strange needle-like outcrops and spiky turrets loomed out of the yellow haze, but it was too grim to get photos.

A storm brewing in Fada

We crossed two escarpments, with more spectacular rock formations, explored an arch formation and eventually passed by the delineated minefield before Fada, now apparently and hopefully cleared. Lots of barrels mark the side of the road and the remains of some blown-up vehicles show what happens to those who stray! The camp in a Fada courtyard belonged to Ismael's grandfather. We all needed to go 'out on the town' to find a quiet place to squat!

27 February: Guelta d'Archei
Fada is a pretty place with lots of trees, mud buildings, arches, some colourful shops, but very little human activity. In a small corner shop selling fabrics from Thailand and Pakistan, we found some notebooks with a map of Africa on the front cover. We bought four as useful souvenirs, since they had no small change!

Leaving Fada our route turned eastwards into a sandy area with several destroyed Libyan tanks, blown up by mines. Further on was an area of fantastic rock pitons of all sorts of shapes and sizes.

Driving through some wonderful canyons with a mix of golden sand and red rocky walls, we arrived at the Guelta d'Archei in time for an early lunch. There is a lot to explore here, but it's hard work trudging through the sand. At the guelta itself herds of camels are drinking at dark pools with the sheer red walls towering above.

Higher up there are said to be some dwarf crocodiles, relics of a past era. But sadly they do not materialise. To get to a second pool, we had to cross a dangerous wall, perching precariously above the black water. On the way back, we all decided to walk through the water and take our chances with the crocodiles and bilharzia! Back at camp, we washed our feet in Eau de Javel (chlorine) kindly provided by Marie-France.

At least there's no wind tonight. Yet we still did not sleep, having to listen to the pitiful cries of some goats marooned high on the crags unable to get up or down – doomed to a night on the cliffs.

28 February: Paintings and Rocks
Heading around into a different canyon, the plan is to climb by another trail to the highest crocodile pools. A large pterodactyl-like bird flew overhead in this prehistoric valley where time has stood still for millennia. Sadly the nearest we came to a crocodile were prints in the sand and a rather large turd! Apparently six children and their parents live in this valley, but of course there is no school.

Pascale, our cook, made some delicious hibiscus tea, from dried hibiscus flowers bought in N'djamena market. They have to be placed in cold water and then simmered for ten minutes 'to kill the microbes'.

The heavily dissected Ennedi plateau has some truly superlative scenery. Massive rock arches, fabulous rock formations and prehistoric paintings are its greatest attractions. It made the long drives earlier worthwhile. Our campsite was definitely the best yet – paintings and rock formations to defy description. One huge painting of a cow, on the roof of a cave was 5ft long. And there were lots of dancing women, goats and giraffes too.

1 March: Oum Chalouba
Sadly it was time to leave the mountains, but more impressive arches and stunning formations were at Deli, where groups come to trek, there is a freshwater spring in the rocks, surrounded by lurid coloured green mosses. Near the main piste south to Abeche a gazelle shot past our car at high speed, flowing gracefully through the air much closer than any we have seen on previous days here.

Above the Guelta d'Archei

Ennedi rock art

Camping nearby

Siân below a vast Ennedi arch

Siân at Kalait market

Kalait is a dusty settlement on the plains below the plateau. We took a few photos, as the people did not seem to mind and bought a small pot for 50p. Later this item was discovered, to everyone's amusement, to be used by women to perfume their nether regions!

A man from Cameroon arrives, searching for a lift to Faya, he has been stuck here for two days. He speaks fluent English and French, but is finding it difficult here as most of the people speak only their local dialect and no French. He said he was on an adventure holiday to see the desert. Apparently he sells goats and sheep back in Cameroon, imported from southern Chad.

Another character down on his luck was a tortoise which had been found by a group of young boys. They were mercilessly teasing it, pulling its legs and poking its face, until Thierry took pity on it and bought it for £2. It was fed water from a bottle top, which it lapped up voraciously. Someway out of town at camp among the rocks the tortoise was let loose, and made its way immediately to some grass, which it started to crunch between its teeth.

2 March: near Kouba
From Oum Chalouba the cross-country tracks would lead back to main piste to N'djamena.

At one well, over 100m deep, two herds of camels were noisily taking the waters. The well was so deep that in order to get the water out, a man riding a camel with a rope attached would walk away from the well, pulling the 'bucket' (made from old inner tubes) upwards. Just as we were about to stop for lunch, a group of villagers arrived requesting water. It turned out that one of them is a military deserter from Faya – he walked here across the desert and has been living with these nomads for two months.

At Kouba a group of children, boys of course, became very aggressive, leaning into the car and playing with the indicators and horn. But an old man waving a stick promptly sorted them out.

Meanwhile, the other car had gone off in search of Ismael's third wife, who is 15 years old like the others when he married them, but she was out! Nour told us that he has also 'found' a 15-year-old to be his wife. Ismael and Nour are both Muslim, but Pascale is Christian. He has three young children, but we did not like to ask how old his wife was when he married her. He did say he does not know how old she is now – just twenty-something!

3 March: after Chedra
It's our longest day, retracing our route. In the attractive village of Chedra we stopped for soft drinks, melons and chickens. Marie-France was fully occupied dispensing eye drops to what seemed like every infant in town!

4 March: N'djamena
Our last breakfast – bread from N'djamena again, with our own muesli in Marie-France's Earl Grey tea as usual!

Playing with the GPS, I found Derby and Burton-on-Trent on the Find City page, complete with map. And Chichester and Bognor Regis, and Llandeilo, but not Llandovery. The world is indeed a small place, if it can fit on to this tiny box I hold in my hand! It's a pity the world hasn't come together more with better communications.

Libyan Journey: 2002

This trip was made when the 'evil' dictator was still in his prime of power. To us the country proved peaceful, mellow and welcoming. Whatever undercurrents were present, they were not visible to the casual visitor. At no time were any officials threatening and at no time was there ever a thought of insecurity. The idea of losing one's head was as far away as Pluto. Is this the same country we observe now?

We travelled with a Libyan guide in our Land Rover for the first week, an affable character with a lot of wild stories of adventure. On the itinerary were Tripoli, the Roman remains at Leptis Magnum, Sabratha and then into the hills and west to Nalut and Ghadames. The old city of Ghadames was a special place with narrow alleys, covered streets and small mud mosques painted white.

Later our guide suggested that we took off on our own to the deep south, to Sebha and beyond. He said no one at any checkpost would stop or question us and indeed that was the case. This part of the trip was very exciting, if a little unnerving.

Gaddafi poster in Tripoli

Leptis Magnum remains, Libya

Siân at the old granaries of Kabao

Curiouser and curiouser!
A curious thing happened in the café at Abuzeyan. On our first trip with our guide, no one spoke to us much. On our second visit, going deeper south without a guide, it was totally different. We were feted with free drinks, attempts at conversation and the TV channel was changed from the local soap opera to the Queen Mother's funeral on CNN.

On the Awbari dune & lake circuit, Libya

Dune Lakes Circuit trip

The highlight was undoubtedly a trip organised locally from Awbari into the dune country to the north. Five fabulous gems of lakes sit among vast reddish dunes brushed by palm trees and a few mud brick houses. The driver's skills were sensational, climbing over high dunes and slip-sliding down the other side. Near Awbari were the strange abandoned ruins of Germu, a ghostly collection of mud remains. In Sebha no one questioned our presence and we even visited a travel agent to see what they offered for groups.

Our Land Rover in less difficult terrain

The Libyan Conundrum

The whole ambience of the country gave no hint of fear. It just did not have feel of being a terrorised country, from our perspective. To us, as rank outsiders, Gaddafi seemed an outrageous buffoon, a dangerous one no doubt, but one who at least prevailed over a country that seemingly had no real army. To keep a lot of tribal clans under control undoubtedly took some strong-arm tactics and probably some clever beguiling talk. With a Tuareg private security force, Gaddafi was a canny player until the bitter end.

Since the great western powers have toppled the dictator, chaos has arisen from the apparent harmony of the country when we visited. This interference may have seemed desirable at the time, but the results are certainly not. It was a kneejerk reaction to a problem that may have appeared relatively black and white to an outsider. Was nothing learnt from the arrogant, misguided and disastrous invasion of Iraq? The precedent was already there for all to observe. Interfering in other cultures is not a simple matter of good and evil.

A war in Mali followed. Timbuktu was lost, along with many of its priceless treasures, and the French had to invade another time. The Tuaregs are again in uproar, back to the bad ways of old, people-smuggling in the poor countries of Niger and Mali. The thriving tourist industry of Mali is lost and the fledgling one in Niger a victim to international mismanagement. Tourism was one of the main keys to stability when we visited these countries.

And finally in 2015 the chaos has led to an unprecedented flood of migrants and refugees from Africa.

With the ousting of the dictator Muammar Gaddafi in 2011, all tourist adventure travel to Libya has ceased, probably for a very long time to come.

Libreville or bust: 2003

It has sometimes been said that the planning and anticipation of a trip is more exciting than the actual trip. Well, at times this was certainly partly true during our latest adventure. But what a trip! Each day we asked ourselves, 'Why are we doing this?' as one difficulty after another arose to impede our plans. After the event we can look back with the certain knowledge that it ended safely. Was it successful? At least we came back in one piece. It certainly wasn't comfortable, but it proved every bit as interesting as any other trip. Had Africa, in particular Cameroon and Nigeria, changed much in 25 years? Yes and no. The biggest change was the sheer number of people everywhere. Of course we expected the cities to be much bigger, but the countryside was also much more populated. Finding a place to camp in the savannah, away from villages or cattle herders, was virtually impossible.

This trip evolved rather wildly and completely divergent from the one that we had anticipated in early 2002. Getting visas was no easy task for any long overland, as we were busy working in Nepal from late October until mid-December.

Fortunately the main reason for doing the first part of the trip across the Sahara was because two small groups of interested clients were joining our trips in Algeria and Niger. It was the first season for years that tourists had been able to head for southern Algeria in substantial numbers. This provided us with a good excuse to drive down to Tamanrasset, instead of flying out. Afterwards we would see how far south it was feasible to get from Niger. In order to pre-empt any problems with our ageing vehicle, we decided to spend some money on it. Now, we thought, we should have peace of mind as we crossed the Sahara.

With more than a week to spare before the group was due in Tamanrasset, we thought we had plenty of time, until the gearbox made some disturbing noises on arrival in El Golea, an oasis normally just three days' drive from Tamanrasset. In El Golea a young mechanic suggested we should return to the city of Ghardaia further north, as the facilities there were superior. The next morning we returned to Ghardaia, where rain, a blustery wind and grey skies added to our misery. Puddles of water covered the hotel entrance; we both got wet feet and the vehicle was parked in a compound across the flooded road. And this was the Sahara!

A knight in shining armour came by the next day, in the form of Mustafa, the uncle of our local agent in Tamanrasset. After scouring the town for a garage that would take it on, the vehicle was dispatched to a workshop with a mechanic by the name of Issa. This name, translated, means Jesus; would he be our saviour? The gearbox took four days to fix; parts came from a village near Ghardaia where dozens of old Land Rovers were slowly dying of old age. Each day we wrestled with 'what if' thoughts.

We made reservations on the flight from Ghardaia to Tamanrasset 'just in case'. Should we fly to Tamanrasset? Would the clients arrive in Tamanrasset before us? Would the convoy south to Tamanrasset be on Wednesday as expected? What if?

One thing that was without question was the hospitality of Mustafa and his family. We were invited into their home without question and treated like honoured guests, a surprise opportunity to live in a Muslim household for nearly a week. Nothing was too much trouble for them. We were unable to cash our travellers' cheques in any of the banks, so they paid for our stay and the repairs to the vehicle – we would sort it out with his nephew later. We were treated to local food specialities such as truffles, something that we could not even contemplate in Europe. As a man, Bob was not allowed into the family quarters. We slept and ate in the guestroom near the front door, with Mustafa but never with the female members of his household. Siân was allowed in the kitchen and TV room, and had to do all the laundry, as that is women's work! His wife and daughters did go out to work, wearing long coats and headscarves but otherwise were independent.

Just three days before the arrival of the clients in Tamanrasset, we had left Ghardaia at five in the evening, after a cursory test drive. By midnight we had reached El Golea, where we discovered the convoy was due to leave at dawn for In Salah. This convoy system, which apparently ran twice a week, was said to protect trucks and tourists from insurgency or just plain old-fashioned banditry.

Dawn broke at the check post south of El Golea. It was a bitterly cold morning; the army personnel were drinking black tea and smoking French cigarettes. Breakfast for them appeared to be bread and oranges. We waited. And waited. And waited. Till 4pm. Finally they told us the bad news. Today the convoy was not to operate and still 1000 rough kilometres to go...

Dawn broke again but this time we were in luck. A soldier gave us a bottle of Orangina in exchange for the fresh orange we had given him from Issa's garden. The convoy set off together, but within minutes we were alone in the vast desert; our slow vehicle being hopelessly unable to keep up. In any case there were no police or army in the so-called convoy. Rain was pouring down, so much so that we had the windscreen wipers on all day long. Could this really be the Sahara? So much rain fell on the black rocky Tademait Plateau that there were incredible rivers to negotiate, with torrential water in huge white waves. Later large pink flowers blossomed in wild profusion, taking full advantage of this unexpected manna from heaven.

Floods in the Sahara

At In Salah, the first fuel station was unreachable because of the truck-swallowing mud. The second had a long line of vehicles attempting to fill up. In the old streets children played happily in the swimming pools made by the unusual precipitation; perhaps some of them had never seen such an event before in their short lives. The police would not let us continue – it was too late for us to reach Tamanrasset. So we camped outside the poshest hotel in town. Inside, the lobby was almost a swimming pool too. The ceiling dripped, and the beautiful carpets had all been lifted and moved to a drier part of the building. We watched BBC World briefly before retreating to the inside of our cosy metal box, listening to the 'tap tap tap' on the roof with a comforting sense of security.

We had hoped for a pre-dawn start, but the police did not rise before the sun! Just after 7am we left, for a long 650km drive on potholed tarmac through the hills to Tamanrasset. The Arak gorge, the only settlement for miles around, an army post and fuel station, came and went... at 6pm we stopped for a tin of sausages and mash on the side of the road. A truck stopped to see if we were in trouble, but we nervously prepared to leave as soon as we saw it slow down. Darkness fell as we continued to drive through a spooky moonscape of rocky spires by moonlight.

Finally a twinkle on the horizon heralded the outskirts of Tamanrasset. It was now a large town with shops, streets and lighting, not the sleepy one-sandy-street village of the seventies that Bob remembered. Mohamed was waiting for us with an open-air barbecue in his garden, but we were too tired to appreciate it. 'The flight will arrive tomorrow at 12 noon,' he told us!

We had two groups, one here and one visiting both Niger and Algeria. With the first group, including intrepid trekker Shirley Kent, we travelled up into the Hoggar to the famous hermitage of Père Foucauld, with its magical sunsets and spiritual serenity, then across the desert to Djanet.

Our Land Rover in the Hoggar 2003

But we only just made it. Once again the Land Rover decided to play up; this time it was the rebuilt engine that fell apart. We met Saharan traveller and writer, Chris Scott, at the campsite in Djanet. He later fell off his motorbike and was flown out, fortunately not too badly injured. It was his luckiest escape, as you will hear later!

With a week of trekking here, we had plenty of time to absorb the amazing rock art and the equally incredible rock formations of the Tassili n' Ajjer. Another Mohamed took the errant engine apart and put it back together again while we were away. We decided that someone 'up there' was looking after us.

Our second group was to visit Niger, taking the route from Agadez to Bilma and then north to the Djado area before making a rare border crossing into Algeria and up to Djanet, changing agents at the inconspicuously named Balise 21. After a short trek on the Tassili they too headed across the desert with us back to Tamanrasset.

When will the Iraq invasion begin?
The group was lively; most were connected with the US military. War in Iraq was now not far off and discussions about the prospects were 'interesting', for want of a better word. They seemed to know when it would all start, despite the fact that the rest of the world thought we were waiting for UN approval. Some views were rather extreme to our minds. Anyone who had formerly worked with the CIA clique and knew Oliver North was probably bound to have strong views. We listened intently to their opinions, but refrained from offering more than a token alternative point of view. It would probably not have been wise. They were, after all, paying us to organise their desert adventure, on which hot showers had now become the main priority.

Once they had departed, we were on our own. From Tamanrasset the tarmac road was broken for some miles through the rocky remnants of the Hoggar into plains dotted with isolated craggy outcrops and massive boulders. Later it was open desert, with many confusing pistes to follow. The world here is breaking apart, destroyed by heat and cold, smashed by gripping winds and sandstorms, slowly decomposing to sand.

In the far south of Algeria, close to the border with Niger, we camped among the outcrops of Gara Ecker. On the roof of the vehicle, we slept intermittently. A cool breeze rustled the plastic bag behind our pillows. The stars were amazing. A large eerie shadow was cast across the bedding by a 50m tower of rock erupting vertically from the warm sands. Sentinels of similar rock towers marched across the desert in the black of the night. Suddenly the headlights of the leading car brushed the Land Rover, and a convoy of two vehicles came to a halt some distance away. We peered out from below the bedding, not daring to breathe. Turbaned-headed men got out of the vehicles. Had they seen us? Of course they had. Leaping around and shouting, they seemed to be arguing. Then another car came past even closer. This was it: bandits, kidnappers, smugglers, whoever they were, we were sitting ducks.

After what seemed an eternity, they got back into their 4X4s and drove off into the darkness of the night, leaving us waiting for our hearts to slow down. Perhaps our crummy old Land Rover wasn't worth the bother, or maybe they thought we were locals, as we had endeavoured to make the vehicle look inconspicuous. All the gadgets were hidden inside, with no gleaming sand mats, jacks, jerry cans or shovels stuck on the outside. Whatever the reason, they suddenly left.

We also left at dawn for the Algeria/Niger border post at In Guezzam. At the border a festival was underway, the diesel station was closed and everyone was dressed in their finest robes and colourful chèches. Sheep were nervously corralled in pens awaiting their fate. It was Eid. 'We cannot find the man in charge of the fuel pump, he is with one of his wives. Please come with us to the police station.' So we were invited to partake in the celebrations at the police station. After much delay, large plates appeared loaded with freshly grilled lamb and salad. All we could offer in return was a McVitie's chocolate cake, which went down very well with the local officers. 'What about the men we saw last night?' we asked. 'Oh, they're just smugglers, cigarettes, you know. We don't have any bandits here, you are quite safe.'

From In Guezzam for about 15km we were in no man's land, crossing an empty plain of sand that was almost too deep to avoid getting stuck. Behind us were the low hills and dunes of Algeria, ahead a solitary dot on the horizon that became an ancient tamarisk tree and a broken crumbling wall that might once have been a proper fort.

Assamaka; how the place has grown since 1978 when Bob was last here. Some basic mud houses have been added to the south, away from the tree and the fort. In a small hot mud building we are greeted enthusiastically by the immigration officers. We are in black Africa now and it shows. They are very friendly and the passports and customs are all done very quickly, which is a big surprise as this border used to take a day or two in the late seventies. But how has this bureaucratic transformation taken place?

These fellows have decided that extracting extra fees for services rendered is more profitable than holding tourists up in this windy and flyblown spot for days. And so we are off to Arlit, heading east along a well-defined sandy track across barren plains of sand and plateaux of low rocky escarpments. We see only two other vehicles and hide overnight behind the first vestiges of bush some way from Arlit. It's a peaceful night, tranquil and starry, but we are still not over the shock of the previous night's visitors and don't sleep much. Our vehicle is much higher than any of the bushes round here, so we must get away at first light.

We pass the uranium mine outside Arlit. In town, the festival is still in progress. The man who issues the insurance needs to be found. Since he has four wives, we have to visit any number of houses to see with which of his good ladies he is currently to be found. We then have to wait while the unfortunate sheep's throat is cut in the ritual way. But it all works out well and everyone is happy, except the sheep. Arlit is a strange town, a mudbrick settlement some distance from the uranium mine, but probably not far enough away to avoid contamination. The mine is currently much in the news; it is said to be the source of material for a possible Iraqi dirty bomb. Of course it's heavily guarded and run by French technicians. The only proper road here runs south, and it's thousands of miles to the nearest port.

It's really quite amazing to be back on a fairly smooth tarmac road, and we don't mind paying the road toll. Suddenly we are back with people again, after three days of almost complete solitude in the desert. It's a remarkable contrast, and impossible to say which is best.

Agadez remains a traditional mud city; its proud inhabitants dress in superb traditional Tuareg robes and chèches (turbans). The tall mud and wood-supported minaret of the main mosque dominates the main square, where musicians and hyperactive horse riders are displaying to the crowds as only exuberant Africans can. The campsite just outside town has beautiful shady areas and running water for showers and laundry. What more could one ask? A few other overlanders are here, coming from the west and east, so we all exchange stories and find that one of them had emailed us back in January asking for information on the trip! In town the next day we go to visit Barney and Sophie, our local agents, and find out how they coped with the CIA in the desert. The drivers remember us from three years ago. Sophie invites to stay at her home; how wonderful it is to sit inside a cool house, and sleep in a bed. Small luxuries we take for granted.

South of Agadez the desert begins to give way to arid scrub with thorn trees, isolated herders and fine yellow wiry grasses. Camping before Zinder we witness a stunning sunset, a giant red disc sinking into a hazy horizon, with distant straw-roofed conical village huts silhouetted against the fading red glow. A solitary herder passes by as we finish off our tinned chicken and Smash potato.

The dusty town of Zinder, Niger

Zinder has an interesting old Hausa quarter with a mud palace and fascinating mud streets. The Hausa architecture is better preserved here than in Nigeria. While looking for somewhere to stay, we are followed by a large 4X4 Land Cruiser. It belongs to Carol and Giles, missionaries from England who live in the town with their three lively children. 'We saw your English number plates and had to find out what on earth you were doing here. Do come and stay with us!'

It turns out the hotel we were looking at is of dubious reputation, and the campsite even worse. Their hospitality is wonderful, and we taste home-made banana bread and mango ice cream among other delicacies. All the children have been educated by Carol, mainly at home by correspondence course. They are so self-disciplined that when we leave at 9.15am on Monday morning, they have already gone to their schoolroom inside the house, and have to be called out to say goodbye! The eldest boy, Sam, will soon have to go to university, so they expect to return to the UK in a year or so.

Kano is a hot place and the campsite offers little shade. Mr. Fix-Everything Hussain has found us and is eager to be of assistance. Nigeria produces oil, but fuel can rarely be bought from fuel stations. There are permanent shortages of fuel, but no shortage of characters able to find some elusive drums. We spend the morning with Hussain, pouring black market drums of diesel into our jerrycans. Need to change money? No problem.

When we obtained our Nigeria visas in London we had to buy US Dollar travellers cheques, so many per day of expected stay in Nigeria, in order to get the visa. So where do you change those dollar cheques in Kano? Well, you don't; banks don't like them and the black marketeers don't like them. Have pounds, will travel in Nigeria. Hussain is getting temporarily rich today, but there is a dearth of travellers in town.

The Sultan's Palace, Kano, Nigeria

This is not surprising, for the country is on the brink of elections, where there have been riots between Muslims and Christians in nearby Kaduna and it's too hot to think. Besides that, there are hundreds of road checkposts. While most of the officers are very friendly and jolly to meet, 'Have a safe journey!', others make it clear that they are not overpaid. Hussain has taken over the job previously done by Mohammed, who has retired. There are only one or two overland trucks a year now. A brief visit to the Sultan's Palace, the dye pits and the old mosque are all we can manage before overheated lethargy sets in after lunch.

We are off east to Cameroon now. Close to the border, riots have been going on between the nomadic herders and the farmers. It's the usual story about land use and water supplies; tempers are frayed. It is the age-old problem of the Sahel. Those who try to live by growing crops and those whose long-horned cattle need space. From Niger to Sudan, the Sahel cannot provide a living for the two groups. We head south from Maroua to Garoua, luxuriate in an air-conditioned room, then head into thicker forest to Ngaoundere. These days northern Cameroon is a risky place, as militants hide away in the hills of Nigeria close by.

We hit a shocking road near the border with the Central African Republic. That country is always having rebellions and is not safe for travellers. Suddenly we have a brand new highway to ourselves; for over 150km we are in paradise, a road with no traffic, and why is that?

Northern Cameroon lunch stop

The tarmac ends abruptly; that's why there is no traffic. We are now on the most bone-shaking corrugated road in all of Africa. Ten miles an hour is our maximum speed for most of the next day. A few logging trucks go by, but little else. We camp in an old quarry; unseen animals and nocturnal noises keep us awake and then at 3am the rain starts. We are sleeping in the open on the roof, so the first drops to fall on our faces are a rude awakening. The road is likely to be shut by a so-called 'Barrière de Pluie' if it rains too much, and our bedding will be like a sponge if we stay here more than a few seconds, so we get moving before dawn.

In a dank, overgrown zone of forest the vehicle comes to a sudden, shocking halt. It is still dark, our torch battery fails and the rain has made the roadside a quagmire of thick sticky mud that sticks so vehemently to the soles of our shoes that we get taller each time we venture out. It's the low point of the trip.

By dawn the rain has stopped and we desperately try to find the cause of this breakdown. The fuel is not getting through, perhaps. After a relaxing cup of tea things seem better, a wire on the injector pump has shaken loose, cutting off the ignition. With much relief we are on our way to Yaounde, where we celebrate our arrival by having an air-conditioned room. Dollar cheques are not wanted in Cameroon either, and our last Euro cash is gone. Yaounde is one big round of trying to change money.

And then the traffic police stop us. We have inadvertently driven down a road that is forbidden. Though unmarked, it is the President's private strip of road in the middle of the capital, and no amount of remonstrating about this innocent action will let us off the hook. Perhaps an 'on-the-spot fine' will sort it out. Yes, the deal is done, £1 changes hands and we head south for Gabon.

Some of the people living along the roadside seem very different from the usual tribal peoples. They have a physical appearance somewhat more akin to primitive aboriginal. They wield large machetes as they emerge from the thick jungle and forest. We are into the equatorial rainforests with trees 100 feet high. The humidity is heavy, the air is still and the road unexpectedly good. We pass a small town where Orange mobile phone adverts are much in evidence. Again, though, there is little traffic and again the road suddenly deteriorates to a narrow muddy lane with large watery holes and rotten log bridges. Apart from the mobile phones, this is Africa as it was in the late seventies. The border is marked by a ferry, but the people are not asking for donations and it is all quite easy. We camp outside the Customs house; it's too hot and humid to sleep much. This is seriously knackering.

Morning comes and we cross the river to Gabon. Gabon too has its good bits of road and its strange aboriginal villages. The pygmies are not far from here, but we don't see any. The road becomes a rutted dirt highway through the magnificent trees. It's being widened, but little work is going on. Camping is a problem; there are no places off the side of the road. At one place we stop to cook but a local comes by and suggests that camping is not a good idea here, too many monkeys or elephants, but maybe it's too many unsavoury villagers. We don't really know. Normally one would expect nothing but kindness and curiosity from the rural populace, but this is a new country for us and it's a bit forbidding.

Rainforest in Gabon

We drive on and on and on into the dark night; the road brushes through thick bamboo forest. After several hours some lights appear; we are in Ndjole, a town on the banks of the wide Oogoue river. We can't believe our eyes as we pull up outside the Auberge. The house could be anywhere along the banks of the Loire in France, except it's hot, muggy and the cicadas are deafening. Inside we are greeted by Madame la Proprietaire and some expat French guests. We have a beer and find ourselves discussing the state of the country with the expats. One is en route to Franceville on business; the other is out for a break from the capital Libreville. Our room transports us to France, to a small auberge, a little tatty, a bit smoky and with rather tacky curtains. It could be in Amboise or Blois. The river flows serenely by, but it's a muddy colour and the far bank has forest lining the bank, not a car park. For all that we have a good night and feel refreshed for the morrow. The Equator is marked by an elaborate sign and then we head on to Libreville and our ultimate destination.

Libreville is a large town with some newer buildings in the main area, but much of the sprawl is shanty-like, tin-roofed houses and bustling markets. It's shockingly humid and we have to find a good place to park. The hotel at the sailing club was the place recommended by the expat at the auberge in Ndjolé, but his budget is obviously far different from ours. Nevertheless some maintenance has to be done and the room is air-conditioned, so the budget just has to be blown! While Bob is underneath fixing some of the shock absorber bushes, a young man comes to tell Siân, 'I'd rather have AIDS then Ebola. At least with AIDS you can survive for a while, but Ebola will kill you within a week.' Currently there is an outbreak of the terrible Ebola virus in Congo, the next country on our route south if we are mad enough to continue.

Libreville markets, Gabon

As it is the rainy season, the beaches to the north are rather grey and empty. Some expats are enjoying a good French lunch, but there is little to keep us here. We head a bit further south to Lambarene, where the tarmac road ends. The setting is beautiful, with the wide branches of the Ogooué River creating a pleasant waterfront. The main attraction in town, apart from the riverside bar at the hotel, is the hospital built by Albert Schweitzer. His original hospital is now a museum and full of his instruments, artefacts and memorabilia. A new mission-based hospital has been built next to the old one and now has quite good facilities.

Asking for a visa to Equatorial Guinea

After a couple of hours at the Equatorial Guinea border, our visa request is turned down. We cannot think of continuing south, with the Ebola virus just 300km south and the two Congos both suffering from armed rebellious groups. It's a crazy trip and there is nothing for it but to return north. Our route is different, though.

We head back to Yaounde, across to Douala, and up to the cool bliss of Buea on the slopes of Mt Cameroon. Then back into fuel-starved Nigeria, loaded with expensive Cameroon diesel, and up to Abuja. In Kano Hussain is waiting to greet us at the campsite, then it's back to Zinder and the missionaries' house. They are sleeping when we arrive at 3pm, but their guardian lets us into the compound and welcomes us like old friends. Their youngest son recognises us when he peers out of the door, and he goes to tell his mother, 'I think Siân and Bob are back.' He's only seven and I'm amazed he remembers our names. It's only three weeks since we left, but that's a long time when you're only seven years old.

In Agadez we stay again in Sophie's large airy house not far from the central market. We are about to leave when they say, 'You must see the Bianou festival; it's on today and tomorrow.' So we stop to see this amazing blast of colour and fun. Camel riders dressed in lavish costumes compete with each other for the most riotous dress and dance.

Watching the Bianou festival with extra guests

It's time to head north again, which we do with some trepidation. We cross the border and get lost in no man's land between the two checkpoints. The sand is deep and we spend two hours digging out the vehicle before returning nervously to the rogues at Assamaka to get better route directions. It is a timely reminder that driving across deserts is not a task to be undertaken lightly. Getting lost, even on the main piste, is always a possibility. There are no night raiders this time at Gara Ecker.

In Tamanrasset we hear that Mohamed had been on the plane that crashed on take-off at Tamanrasset, but he had arrived on that plane from Ghardaia just before it crashed. He mentions something vague about some missing motorcyclists, but that's all. We pass the burnt-out wreckage of the plane, scattered on both sides of the road as we drive north on the main route to Moulay Hassan and In Salah.

The Iraq war has just begun as we watch the TV in In Salah; there are pictures of the British parliamentary debate about the war on the Arabic Al Jazeera TV station. They play some scenes of mass demonstrations marching through London. Our Algerian hosts are quick to calm our fears;

they can plainly see that it is governments who take actions and that much of the population are not in agreement.

As we cross into Tunisia, we are stunned to hear that thirty or more foreigners travelling in their various private cars and on motorbikes have been kidnapped by an Al Qaeda-linked group just north of the Tassili, some 300km northeast of one of our camps near Moulay Hassan. Evidently we and Chris Scott have had a narrow escape. We are relieved, yet shocked and saddened. At the campsite in Tozeur a group of Germans are hoping to go into Algeria to look for their colleagues and flatmates. Information is patchy as we take the ferry to Genoa.

Soon we are home and the news from Algeria is bad. We are perhaps very lucky not to be spending the summer in a hot desert, held captive by angry men. The trip was very interesting, but one has to seriously wonder if such adventures are becoming more of a risk and less of an experience to be savoured. For us, though, driving across Africa has become almost an addiction.

Hostages

In May and August all but one of the European hostages were released unharmed, except for the traumas and scars resulting from a long period of captivity. Sadly, one woman had died from heatstroke. The British media said little about the whole episode, since there were no British hostages. But at the time of the release of the second group of hostages, we were in Switzerland. Shockingly, just a couple of sentences after mentioning that the Swiss hostages were finally free after nearly eight months, talk began about who was going to pay for it. It was even said that they should not have gone there because it was too dangerous! This at a time when the British Foreign Office had just updated its advice to say it was okay to go to southern Algeria. No other foreign government had foreseen any such event either.

Such is the state of the modern world. For the time being the tentative beginnings of a tourist revival in the Sahara are extinguished.

Soon we are back at work in the Alps for three months with trekking groups.

Central Asia: 2003

Abridged from Siân's diaries.

The surprise of this trip was being able to do it all. Turkmenistan had a rather bad reputation and the fact that all travel had to be organised in advance with a local operator seemed merely to add to the foreboding. Uzbekistan also implemented the same philosophy for the early travellers to its charms. Of course the Soviet Intourist Company had long allowed tourists to take strictly controlled parties. Now the doors were creaking open to more independent travel. We set up the program in advance here with local operators. Beyond that Kazakhstan and Kyrgyzstan seemed only a few miles ahead, with more freedom to move about.

26 September 2003: Istanbul airport
Arrived Istanbul 5.30pm for a long wait. We can't get into the HSBC lounge without an HSBC credit card – I have an HSBC debit card and NatWest credit card, but not the right combination! So we sat in departures till 10pm.

27 September: Ashgabat АШГАБАТ
Our flight landed early in Ashgabat at 3.30am. But the bureaucracy was ludicrous – it took three hours to get out, with nowhere to sit while we were waiting but on the cold tiled floor. We emerged at 6.15am to find our guide, Tatiana, anxiously waiting for us. An American tour group with a Turkish/American leader is doing almost same itinerary as us. We drove to the Nissa Hotel with a fabulous view of mountains – it's dawn now – fabulous room too.

Later Tatiana took us to the city centre via the Presidential Palace, Congress, a statue of President, and the Earthquake Memorial. Generally the city is very quiet and clean with lots of green open spaces. We ate a grotty lunch in the Russian market. Back in the room we needed some rest, but then the cleaner knocked the door for no obvious reason, probably to spy on us. We know that the country is run by the President with an iron fist and kept in check by the secret police, and that no dissent is permitted.

Siân in Central Ashgabat, Turkmenistan

Tolkuchka Market, Ashgabat

28 September: Ashgabat АШГАБАТ

Probably the most interesting place for any tourist near Ashgabat is the Tolkuchka Market. It's a bustling place famed primarily for its carpets with lots of red colours. Hats are another speciality here. The people were friendly and curious. There were no cafés that we'd dare to have tea in! Or toilets, despite thousands of people!

One of the most bizarre sights in the capital is the central tower with a revolving golden statue of the president on top watching all below. Some high-speed lifts shot us up to the top viewing area. The impressive-looking civic buildings give a hint of the new oil wealth and there was also the Istanbul mosque to visit, but it was firmly locked. Around 5.30pm the surprise was that the muezzin was calling out… a recording. Across the whole town we saw few indications of religion, with hardly anyone wearing a headscarf. The newly independent states of Central Asia are perhaps understandably a little nervous of any religious opposition, with the horrors of Chechnya only too recent.

29 September: Dashoguz ДАШОГУЗ

Alarm at 5.50am – ugh…. – but the breakfast room opened only at 6.30am.

Finally we found Tatiana, 'How do you want to go to Dashoguz?' 'We have already paid for a private taxi.' 'Yes, but I can give you the money back if you want to go by minibus, much cheaper. Let's go and see.' Looking at the minibuses on offer, it was clear we were not taking any of those across the Karakum, the black desert of death. We chose a car with reasonably good tyres, a red Lada whose driver wore a furry Russian hat. Tatiana noted his registration number and off we went. Within a few minutes we had left the swish city with its fancy gold buildings and wide avenues. It disappeared quickly into the distance as we shot off along the highway – dual carriageway at first but not for long.

The road became worse and worse, the scenery bleaker by the minute. We stopped at 12.00 noon and the drivers disappeared into a yurt to eat shashlik, and a melon which one of the policemen at the checkposts had given him. It had been taken from a truck driver as a bribe while we sat and watched! We shared the water melon too as we sat outside in this tiny settlement – a yurt, a well, a small concrete building and a few other small scattered buildings. Further north we passed a canal, then more and more water, irrigation for the cotton-growing fields. The ride seemed endless and it was 7pm by the time we reached the bright lights of Dashoguz. The only other westerners in the place were updating the Lonely Planet guide. They are often the only ones in such places!

Khiva city walls, Uzbekistan

30 September: Khiva КИВА
There is no sign of our ride to the border or Kemal the hotel owner. They all have gold watches, but no one knows the time! Later there are no hassles leaving Turkmenistan or entering Uzbekistan and soon we meet our new guide, Viktor. It's just that Viktor couldn't find any petrol for his car.

'Problem with petrol here.' After some time again we are off, Viktor has found a man with a black market supply in some plastic jerrycans. The stylish old car is a great-looking Lada with sofa-like seats and soon we are in Khiva.

Unfortunately, with all the petrol in the back in leaking plastic containers, there is rather an unpleasant aroma en route. In the wonderful old city we sit down below a large green minaret for a lunch of borscht soup and bread with tea costing just $2. Another Lonely Planet writer sits next to us and again there are few other travellers to be seen. We are free to walk around on our own. Viktor is no doubt seeking out more petrol and some vodka for his evening off. The Kukhna Ark was our first port of call, then we wandered round the old city streets gazing at the amazing blue-tiled arches, mud buildings and mosques, as well as an underground water tank.

1 October: Khiva КИВА
Over a good breakfast of eggs we are joined by an American travelling alone with a guide – he may be a carpet trader perhaps. He was in Afghanistan recently, just after the war, and said a bed costs $60 in a rough hotel in Kabul. And the cost of a roast chicken shot up from $3.00 to $7.50 in a month because of all the foreign aid workers.

Panoramic view of Khiva

The 118 horrendously steep dark steps of the Islom Huja Minaret took some doing but we were rewarded with fine views. We met a Pakistani woman working for the UN on drugs, criminals and terrorism. She had flown here for the day from Tashkent with her Austrian husband, who was visiting from Vienna.

After seeing the South Gate we went to the carpet workshop, and had a guided tour by Christopher Aslan, a volunteer working for Operation Mercy in association with UNESCO. They are reviving the lost art of silk carpet making. It was all very interesting and they had some beautiful designs. The workers earn $50 per month. Apparently a teacher only earns $10–15 per month. He explained how the silk worms are grown in people's living rooms, eating mulberry leaves, then the cocoons are taken and the silk is washed, treated and dyed in many large pots and finally woven into carpets. Chris was delighted to get our Time magazine from the plane and commented that he is probably the only Englishman who hasn't had enough of the Hutton Enquiry into the Iraq War.

2 October: Bukhara БУХОРО
Time to leave Khiva – it's been a nice peaceful place to visit, with lots of interest.

We had breakfast with two Aussies who were travelling around Central Asia on their own Australian motorbike, which they had flown into Schiphol airport in Amsterdam because that was much cheaper than shipping it to England. They'd left their bike in Bukhara and planned to fly back to Tashkent to collect their Turkmen visas in 10 days.

We met Viktor again and were on our way, mostly at a breathtaking speed of 120kph, dodging potholes and slowing whenever his police detector went 'Beep! Beep! Beep!' We crossed the wide Amu Darya River on a pontoon bridge, which failed to float as the river was dry from the over-production of cotton here. The desert was pretty boring and Viktor had to disappear into the bushes again in search of contraband petrol whenever we were close to the Turkmenistan border just across the river. In Bukhara our accommodation was at the fabulous, fabulous Hotel Sasha and Son. The room had wonderful paintings and plasterwork and BBC TV.

Ancient Bukhara square

3 October: Bukhara БУХОРО

Nepal featured on Asia Today (BBC) with a report about the growing Maoist insurgency. The report catalogued some murders of village men by the army; it is not good.

Bukhara was not disappointing. We visited the Char Minar, the Kalon Minaret with superb blue and purple mosaics, and the streets of the old city. The historic Ark holds some interest, for this is where the two British spies (in the Great Game), Connolly and Stodard, were executed by the Amir of Bukhara.

There were many brightly tiled medressas, narrow sandy streets and carpet sellers in restored markets, but no food to buy anywhere, only a shop selling beer and soft drinks. There was a lot of restoration, but the city still seemed lifeless, devoid of many people in comparison with other places we've been.

There was a very interesting BBC programme about Stonehenge, but it seemed odd to be in Bukhara learning about English heritage! We also met an old American couple. He was 80 and she was young at 72! They were travelling alone with just a driver. Amazing.

4 October: Bukhara БУХОРО

Took a taxi to Samani Park – wandered back slowly via lots of monuments and found an atmospheric tea/carpet shop in one of the medressas. On the TV Bush and Blair are trying to convince themselves and the world that the war in Iraq was justified, despite the Interim Weapons Report saying no weapons have been found. The breaking news is about a suicide bomber in Haifa.

5 October: Samarkand САМАРКАНД

We awake to find Israel has already retaliated. Will it ever stop? And according to reports, UK foreign secretary Robin Cook claims that Tony Blair knew there were no weapons of mass destruction in March when he went to war. (Cook later died suddenly on a remote hilltop in the UK.)

Registan Square, Samarkand

We went to get fresh bread from the bakery before leaving for Samarkand. Endless cotton fields and mulberry bushes along the roadside are again testament to some serious ecological disasters in the country. From the hotel, miles out of town, Viktor came and took us to the Bibi Khanym

Mosque. The mosque is astonishing; so colossal it's unbelievable. Apparently the structure almost crumbled under its own weight.

Outside town there is a hillside covered with graves called the Shahi Zinda complex; it has some elaborate blue arches, facades and tombs. The main highlight amongst many in Samarkand is Registan Square. Each of the three different medressas around the square is quite sensational for the colour and artistry of the tile work. Inside the northern medressa is an indescribably breathtaking blue and gold archway.

It's a heck of a long walk back to the hotel as Viktor has gone walk-about and there's not much food about.

6 October: Samarkand САМАРКАНД
At least the breakfast was good – it seems to be the only food we can get around here. We were very hungry after yesterday's lack of food.

Uzbek ladies in Samarkand

The Ulugbekh's Observatory is an amazing place built around 1400AD. Apparently Ulugbekh died when his son chopped his head off: a nice way to treat your parents! At the market Viktor dropped us off to make our own way back. The weather was awful, so much worse than yesterday.

In the afternoon a miracle happened. The skies cleared fast so we dashed off to find a taxi to take us back to the Registan. In the luminescent sunlight, the Registan Square was magical. Fantastique! All worthwhile! The fabulous Guri Amir mausoleum tomb of Ulugbekh and his family has more stunning gold and blue. Back at the hotel we finally opened our emergency rations: tinned salmon in tomato and basil sauce, eaten out of a glass with a plastic spoon. It made the bread taste quite edible!

7 October: Tashkent ТОШКЕНТ
The drive to Tashkent took hours longer than expected. The old Soviet motorway route goes through Kazakhstan, so we had a 100km detour via Gulistan. Crazy! Viktor kept stopping to buy

apples and then some large, expensive smoked fish. Eventually we reached Tashkent around 1.45pm. The office of our travel agent Dolores Tours was in the Hotel Grand Orzu, as they owned it! Finally we could pay for our trip with the money that had been stuffed uncomfortably around our waists and in an elastic bandage on Bob's leg for so long.

There was definitely no direct bus to Bishkek, so we decided to leave tomorrow and give ourselves an extra day to get there. Had mushroom omelette for lunch, then went on a whistle-stop city tour with Viktor. Wide streets, lot of parks, some very stylish new buildings, just a few Soviet blocks. Saw Earthquake Memorial and Parliament buildings.

8 October: Taraz, Kazakhstan ТАРАЗ
After the feast of breakfast, Viktor didn't arrive till 9.15am. Just the one day we really needed to get away early. But never mind… All was forgiven when we reached the chaotic border and he talked his way past most of the guards and assisted us through a gate, fighting a path through an unruly mob. But then suddenly we were on our own in Kazakhstan.

There were no buses at all, just lots of cars, mostly with totally threadbare tyres. We found a little money exchange booth down the road, then returned to the first taxi driver who offered us a lift to Shymkent and Taraz. His car had new tyres and headrests on the back seats, so was an obvious choice. We said OK to Shymkent and decided to test his driving skills before committing to the longer journey. Along the way he kept trying to communicate – he had been an engineer in charge of a large project in Damascus, Syria for three years. But now engineers have no money here and he's better off driving a taxi. In any case he was a nice man and after a while in his car we decided to go all the way to Taraz.

It was a great ride – beautiful sunshine, shades of gold/beige/brown on the rolling agricultural hills, with a stunning backdrop of white mountain peaks.

By 2.30pm we had reached Taraz, and he put us in a local taxi to go to the Hotel Zhambhyl. In the hotel, the receptionist tried hard to explain, without success, why we could not stay there. Oh no, we should have stayed with Abdul Rakhim in his comfortable car!

Soviet styles in Taraz, Kazakhstan

Another taxi outside seemed surprised we couldn't stay there, but offered to take us elsewhere. So off we went on a wild goose chase through potholed back streets, wide boulevards and extensive parks – it's a huge city. None of the hotels he took us to would take us, and we were on the verge of going back to the first one and dumping ourselves in the lobby! But then he found the Kohak Ily (or something like that!) and a very friendly woman who gave us a basic room for 1000 Tenge ($7). But he wanted 2000T for himself and we settled on $10, having no more local money.

Off for a walk down our street, we found a busy market and an immaculate new money-change building tacked on to a tacky Soviet block. The woman at a teashop smiled and said, 'Latvia, Lithuania?' Her little boy was learning English at school, but he was very shy and clearly astonished to see real English people in his shop, so he just stood and said 'Goodbye' quietly as we went out.

Thank goodness we'd had a good breakfast... otherwise two consecutive meals of dry bread and peanut butter and strange cheese were our only fuel today. Oh, we also spotted some bananas in the market to add to our restricted diet.

It had been an interesting day but these ex-Soviet states are a bit drab and depressing, yet some people are so friendly and want to help even though we can't communicate. Such is the human spirit.

9 October: Bishkek ВИШКЕК

Up just before 7am with only a dry bread roll and peanut butter for breakfast. We daren't try to find a teashop with an unknown journey ahead.

So, just after 7am we were outside on the roadside – very quiet – looking for a taxi to the bus station. A man then took us across to a minibus for Bishkek; we sat inside and waited and waited. The bus was full of people and possessions; our legs were rather cramped but soon we were moving. The people are an interesting ethnic mix – white Russians in smart fur-trimmed coats, very large Mongolian-eyed women in voluminous flowery print dresses, a man in a suit carrying a briefcase, and us, two scruffy European backpackers!

It wasn't long before we reached the border with Kyrgyzstan, and oh how simple it was. At a supermarket we stocked up with Twix bars to use up some of our Kazakh Tenge, and were then through both sides of the border within ten minutes.

In Bishkek tea was our first priority, then we had to try to phone the travel agent who had arranged the visa, Rajiv of Asia Tours, from a complicated local telephone office. Pay a deposit, dial the number, press three, talk, then pay afterwards. That done, we sat and waited for him to arrive, munching yet more dry bread!

At 2.15pm Anwar, another Indian, arrived, and took us first to the Business School, but while we were there looking at a nondescript room, he received a phone call saying there was an apartment available with television, for only $11.00. Then he mentioned that he is going tomorrow to drive round Lake Issyk Kul and back on Saturday afternoon. Would we like to go? We'll be back, inshallah, in good time for our flight on Sunday morning at 4am... Yes please!!

So there was just time for a quick exploration of the leafy city with no particular sights to miss.

10 October: Karakol КАРАКОЛ

At 8.30am Anwar appeared and we were off on our long drive east. Though still tired, it was an exciting prospect, which kept us awake.

Very soon we could see snowy white peaks close to the road, and it just got better and better. We stopped for lunch at a roadside yurt complex and had soup and shashlik, with tea of course, then continued through the dramatic Shoestring Canyon. Wonderful autumn colours everywhere, green grass and white water. Very exciting!

At the western end of the huge Lake Issyk Kul, after passing the town of Cholpon Ata we turned north up the Semyonovka valley, on a dirt road. It was a long way up this very scenic valley to a wide open plain where a few yurts shared the ground with huge herds of horses, sheep, goats and cows, including some that looked like the Swiss Simmentaler cows.

Here he wanted to find a man who wasn't there – 'maybe two hours, maybe three, he's coming...' so we wandered around in the cold but bright sunshine taking photos. After a while he was found– they need him to arrange 40 horses for an Israeli group next week. It was good to have a break from driving and explore this pretty forested valley.

From here we rejoined the main road and continued eastwards to the far end of the lake, then turned south to Karakol. We enjoyed inspiring views of the Tien Shan range, a great white band stretching into infinity. After a brief stop to inspect the centrally-heated yurt where we would be spending the night, we went on to the Kalinka Traktir restaurant for a well-deserved dinner and a Russian Siberian King beer. On a table opposite us was a group of young Kyrgyz people, their faces just like Tibetans or Nepalese Sherpas, and their smiling attitudes identical too!

11 October: Bishkek Airport ВИШКЕК
Leaving our yurt after a cold night on a hard bed, we scoured the village for somewhere to have breakfast. Eventually we found a place for omelette, tea and bread.

The canyon of Jeti Oguz is a truly fabulous picture, with eroded deep red rock shapes, including the famous broken heart, a massive heart-shaped rock, perhaps 30m high, split down the middle. We drove up well beyond this point, up a steep-sided gorge, crossing a white water river three times to reach a wide open plain. Snowy peaks, green fir and brilliant red and orange leaves made a wonderful picture, with a few yurts for good measure.

We stopped for lunch near some dazzling red bushes and a sandy beach. Huge pebbles below the surface and steeply shelving ground meant lots of waves lapping against the shore. For lunch we had fresh trout from the lake – a deep pink colour like the rocks round here, several of which found their way into my pockets! I didn't believe Anwar when he first said we were having fish from the lake, but when it came it was a really special treat which we would have been sad to miss.

High in the Tien Shan, Kyrgyzstan

The village of Bokonbayeva has a different sort of appeal. We were in search of Saginbai, the Eagle Man. He has clients including members of the Queen's household (our Queen, that is) and various Scottish lords, yet no one here seemed to know where his house was! When we eventually found it, he was temporarily out, so his wife showed us his four falcons and invited us to tea and to inspect the house. They had a magnificent collection of furs and horns and photographs.

Saginbai, 'the Eagle Man'

Saginbai himself arrived later as we were tucking into homemade jams of home-grown fruits and then a delicious vegetable stew. The trouble was, we were still full of fresh trout from lunch. Two weeks of starvation and then this feast! He gave us a demonstration of the magnificent birds, weighing 8kg and with a wingspan of around 2m. He has caught them all in the wild, is obviously very skilled at their handling and very fond of them, as they are of him. He affectionately stroked their necks while they almost 'purred' in response. Live pigeons are kept nearby for when they get hungry and are not hunting.

Jeti Oguz outcrops

At 5.30pm we left, still a long, long way from the airport. It was dark soon after we left the end of the lake and were hurtling along an unlit and unmarked highway at quite a pace. With only a brief cigarette stop for the driver and loo stop, we still didn't reach the outskirts of Bishkek until 9.30pm. Then we went to a freezing outdoor shashlik stand for slimy chicken and more tea, before reaching the airport at 11pm.

The airport building was also very cold, and still we had five more hours to pass... The seats are of thin metal and were too cold to sit on for long. The only heated place seemed to be the Immigration men's office – even the woman checking passports had to sit in a greatcoat as she inspected us with great severity. It was 3am when the American group we had last seen in Khiva arrived. They must have known something! Just in time for the flight though.

12 October: Chichester
Incredibly at Istanbul airport while waiting for our connection we saw two young women slumped on a seat opposite. Then one of them came over and said, 'Hello, Siân and Bob, isn't it?' With a flash of recognition we realised it was Tanya, whom we had last seen in the Rum Doodle on 31 December 1999. She used to work for Keith Miller/Explore, was now working for a charity in Devon and was returning from a three-week holiday in Mongolia at an eagle festival in the wilds of nowhere. The rest of the journey was uneventful and we were pleased to see Beryl and Tony waiting for us at Heathrow... and looking forward to her home cooking!

London to Cape Town: 2004

Déjà-vu
It was déjà-vu again; January 2004, and yet again we were planning to drive across Africa. After our rather curtailed trip to Gabon last year, it finally looked as if there might be a way to do the Grand Overland across Africa. It all revolved around getting a Sudan visa and being able to get from Aswan to Wadi Halfa by ferry.

First we obtained the visas for Ethiopia and Libya, paid for a ferry from Genoa to Tunis and even arranged an expensive carnet, including Egypt. Already in at the deep end, we obtained the Egyptian visa in London and sent over three hundred pounds for the Aswan ferry to an Egyptian travel agent with whom we had had only cursory e-mail dealings about future desert programmes. Now that is crazy! For good measure we e-mailed the Djibouti Embassy in Paris to be allowed to collect a pre-arranged visa in Paris en route.

Such planning but then, after years of experience, we knew that having as much paperwork as possible in order before the trip would mean less time and costs wasted en route. And time was what we really did not have enough of, with only four months before the start of our work in the Alps in June. And what about the Land Rover at the end; could it be shipped back in time before the summer? So many unknowns.

We waited and waited all through the middle of January; still no Sudan visa. It was said that the visa could be obtained in Cairo, but we had so little time we needed to be in Aswan for a ferry on 16 February, or the rest of the trip would be too rushed. In the end we had to book our ferry from England to France and buy all the food supplies before getting our Sudan visas. We bought about sixty tinned meals from Sainsbury's, breakfast cereals for three months and pies, cakes and desserts for about three weeks. Departure day was almost upon us, 28 January, with just two and a half days to get to Genoa. We would go anyway and hope...

Then on the evening of 26 January the Sudanese granted the visa; we could collect it tomorrow. So the trip was on. Of course we had had doubts all along, and not just about visas, for undertaking a big trip like this involves much more. Security is the byword now in all the media coverage of world events, and Africa has more than its fair share of these concerns. Do we really need to do this trip? Why should we take all these risks; breaking down in some awful spot in hostile country, kidnappings, robbery, theft of the vehicle, murder, tribal animosities, banditry and the notoriously unsafe road from Ethiopia to Kenya. Then there is malaria, dengue fever, stomach bugs, sweltering heat, dust and even cold in Ethiopia. Why bother? Why put up with the discomfort of living in a 6x4-foot tin box stuck on an ancient Land Rover chassis with an underpowered engine and a gearbox that was supposedly reconditioned last year. So many nagging doubts! On top of all that we both have ageing parents, and one cannot exactly drop everything and fly back from most of the route to Cape Town.

27 January: London
We paid the congestion charge by phone, drove up to London, and, while Siân went to the embassy to get the visas, I drove around increasingly larger circles in central London as every time she appeared from the embassy, the passports were not ready, and the traffic warden was on the prowl.

On top of this my mother was still in hospital after a hip operation that had been done long before. Recovery was a bit slower than expected, so every day we were visiting the hospital in Midhurst.

28 January
We have still not packed the Land Rover, the food is thrown in the back and the final spare parts are still not in lockers, and there is only two hours to get it done and off to Newhaven. We visit the hospital early and get stuck in Sainsbury's in Chichester, having forgotten in our last-minute panic to buy any instant mashed potatoes, a vital ingredient in days to come where even carrots and tomatoes cannot be found. The ferry is at 7pm and it's nearly 5.30pm when we leave.

On the way to Newhaven a terrific snow blizzard erupts. It's the worst blizzard we've seen for years and we arrive late for the ferry. With high winds the ferry has been delayed for four hours. It's bloody cold. The terminal is gloomy and dismal; the only heat comes from the hand dryers in the toilets. We sit eating our sandwiches in a freezing cold vehicle, occasionally starting the engine to get some heat. The heater isn't exactly powerful. The longer I (Bob) sit here, the more I wonder why I am doing this. After all, I've done this sort of thing before. I mean, I've done London to Cape Town and I've done London to Jo'burg before, and things were a lot easier, safer, and then I had twenty clients to push the vehicle or give some help. I'm too old for this caper. But we still sit there; we just can't turn around – we are compelled to go on.

29 January
We have grabbed a few hours' rest on the freezing docks at Dieppe. The ferry came in at 11pm and we sailed to Dieppe. It's a sunny morning and the snow lies all about us, deep and crisp and even deeper in places. In fact it's so deep, there's hardly any traffic, and we are slipping along at less than 25 miles an hour. We are never going to make Genoa by Saturday morning. All the way to Paris we have icy cold wintry conditions, then things begin to get a little warmer.

We head into Paris; it's a Thursday morning and, with a map obtained off the Internet, it's surprisingly easy to find the Djibouti Embassy, where they are very helpful indeed. We are given a temporary permit to park outside the embassy and Siân is busy inside learning all about the delights of the country. Outside, I am still rearranging the food and spares and all the clothes that were thrown in the back. We have visas for Djibouti in an hour and the embassy staff can't quite believe we are intending to drive all the way there. Nor can we! We are off on our way to a Formule 1 hotel at Auxerre. It's still bloody cold.

30 January
We make it to the Alps and, of course, it's bitingly cold. In Chamonix we stop and accidentally bump into an old acquaintance. He is skiing and the slopes are in good shape. We say we are on our way to Cape Town; it doesn't seem real and anyway we are too cold to think about that much as we have a quick brew. The trouble with doing a trip across Africa, or any other continent, is that before you've done it you don't know if you will succeed, and when you have done it, it seems to be something that happened last week. You never really sit down and think, 'Wow, we did that!' And anyway it doesn't mean much to anyone else. 'Oh, really?'

We headed off through the Mont Blanc tunnel, it being -10°C in Chamonix, and later camped near Ivrea, a delightful place just out of the mountains. We made the ferry to Tunisia and not long after found ourselves back in the same olive tree-shaded campground just outside Hammamet, which we'd been in just a year earlier en route to Algeria.

Wadi Halfa or bust
Heading quickly south, we camped near the Libyan border with thirty French campervans on a trip to Libya. We planned an early start; we certainly didn't want to be behind that lot at the border. The

best laid plans! By the time we were up, all but four campervans had already gone; so much for the border. But as luck would have it, we scooted across the border into Libya, bypassing the column of campers. Here we met our compulsory Libyan guide from the company Africa Tours, based in the south of Libya.

Sahel was also from the south, tracing his black parentage back to the southern Sahara. With a deadline looming in Aswan, we rushed off across Libya as fast as possible, to the obvious dismay of our keen and interested guide. In three long days we made Tobruk and stopped briefly at the War Cemeteries, before reaching the border and saying our farewells to Sahel.

The Egyptian border was a place of considerable confusion with regard to paperwork, and of course a vast sum amount of money departed from our wallets to Customs for this bit of paper and that, and some other expenses no doubt. 'Welcome to Egypt.'

By the time all the scratching of the Land Rover chassis to find the number, combing the engine and other meticulously carried out tasks were completed, it was well past dark. We drove to the next village and parked in the car park for dinner. A can of mince and spaghetti followed by a Bakewell tart each sufficed. A curious young soldier appeared mid-way through this plethora of delights, but said nothing and simply smiled at us. He left when we climbed into our tin box and shut the door.

Four hours later, at one o'clock in the morning, there was an aggressive knock on the door. 'Open! Open!' 'Just a minute, please,' we said, and dressed hurriedly. More soldiers were gathered outside wishing to see us. 'What are you doing? You cannot stay here. You must go! Welcome to Egypt.' Their 'Welcome to Egypt' had an empty ring to it. It became apparent that we could not possibly stay there.

Do you think we could find a place to camp after that? The desert was fenced off to the public in most of Egypt by lines of raised rubble left from road building, and the only places left were old quarries and fuel stations. We found a track off to the side on which to camp, but by that time we were halfway to Alexandria. Things were no better the next day, with a continuous stream of beachfront developments all the way from El Alamein to Alexandria itself. We finally found a place halfway between Alexandria and Cairo, called the Fisherman's Village. It was a delightful hotel complex beside some lakes in the mostly irrigated desert. This road, called the desert highway, was once in the desert when Bob took it in 1980, but now! Where has the desert gone? It's receding westwards towards Libya at a great pace.

Cairo
Traffic, traffic and more traffic! We came into town at the back of the Pyramids, where new luxury hotels have taken root in the desert. Le Meridien was the first hotel we found where we could park the Land Rover. With the explosion of Cairo, its traffic and its construction, all the tourist hotels have moved here, as getting across Cairo is impossible in less than a day. Such is progress. We parked and called our travel agent, Solara Travel, half expecting no answer. But, 'Welcome to Egypt.' Amazingly, our man Tarek was in and ready to see us; we waited at the hotel and soon enough contact was made.

Tarek took us to a campsite and arranged to meet us the next day at a papyrus factory on the Saqqara road to sort out the finances of the ferry. How very Egyptian! We spent a delightful morning actually doing the sights, the Pyramids at Saqqara being our prime destination for the morning. Of course we got lost trying to drive there, but in the bustling back streets everyone seemed keen to help us; some did not even ask for any bakshish, very un-Egyptian! Back on the Saqquara road, we waited at the papyrus factory for our main man.

It now appeared that, although our booking had been made with the Nile Navigation Company in Aswan, the agent had been unable to pay the money over. A special rule applied to this ferry, unlike any other in the world. Surprisingly, the money that we had sent at great expense to Cairo from our bank in Chichester was now given back to us in cash dollars, in order for us to pay the money directly in Aswan ourselves. Even more surprisingly, we actually got back one dollar more than we had sent, such are exchange rate quirks. Tarek was an amazingly honest Egyptian, whom we had the greatest pleasure to deal with. Solara Travel cannot be rated highly enough. Because of his honesty, we felt compelled to buy a painting at the factory, in order to give him and his colleagues some well-deserved profits.

It was time to get moving. We took an obscure route down the west side of the Nile, getting lost before Beni Suef. Buses and minibuses poured out choking black exhaust fumes and donkey carts dropped out equally pongy waste. We took a route across the desert to the Red Sea and again there were no places to just pull off and camp. A million square miles of empty desert, but not a single place to camp in the wild. At St. Anthony's Monastery we were refused permission to camp outside the main gates in the desert. So we drove on. We thought about camping out in the desert, but when we saw a strange man wandering in the dark, alone, miles from anywhere, we thought better of it. We tried two posh resort hotels along the coast, but no, it was said that the police did not allow one to camp in the car parks or on adjacent waste ground. They would call round at 2am to check hotel records and ascertain who was staying where. It seems that one cannot just pull off into the deserts of Egypt and camp. This really is a police state but at least it's safe!

Finally at St. Paul's Monastery we pulled off the road into what appeared to be a small compound. Instantly we found four machine guns pointing down our throats. Too scared to be terrified, Siân leaned out of the window and smiled into the darkness. 'We are tourists,' she said, hoping it did not sound like, 'We are terrorists.' The young boys handling the guns told us to get out off their compound and go through the gates to the monastery grounds, where we parked on the side of the road next to some small coaches below the monastery. Why this sudden change of rules? Next morning we discovered that a big festival was taking place at the monastery, with thousands of Coptic Christians celebrating. Everyone was sprawled about in their thousands, camping wild at the site.

The coastal resort of Hurghada is being taken over by Europeans buying second homes or even first homes, with their lower pensions escaping higher costs and cold winter weather in Europe. Constructions are mushrooming all along the coast, but it's not unpleasant. We parked our Land Rover outside a German couple's new home; we met them at a chemist's shop. We stayed in a hotel down the road, a wonderful place for just £6.

A terrible wind blew up the next night in a proper campsite at Safaga, which marred our first real opportunity to enjoy the peaceful surroundings. There is always something to complain about! A French overland truck drew up later, 'doing' the Nile; at least that added to the evening's entertainment.

By now we were into mending flat tyres almost every day. It's not that we actually had any punctures through the tyres; we seemed to be suffering from a dodgy batch of useless tubes bought a year earlier in England. Well, we are keeping a lot of tyre repairers busy most days!

It's a well-kept secret that this country is perhaps not as safe as the glossy brochures would have one believe. Still you can't blame the authorities for keeping some things quiet. A lot of people need the tourist industry. For security there are daily, early morning compulsory convoys. It's a bit of a hassle, to say the least; if your vehicle can't go faster than 80kph, you're mincemeat with the cops.

By the time we reached the halfway tea stop along the road from Safaga to Qena, we were thirty minutes behind the heavily armed convoy. It's a staggering sight. Fifty or more large turbo-charged, air-conditioned coaches travelling at high speed across the desert every day of the season, taking its human cargo of day-trippers to see the sights of Luxor and beyond. Only one solitary vehicle trailed behind this modern caravan across the desert. Our escorts either feigned total disgust with us or had a good laugh as they sat inches from the rear of the Land Rover.

The last part of the journey follows a large road beside the east bank of the Nile, and this road was completely devoid of local traffic for the period of the convoy. Today the poor local fedayeen waited a very long time to be set free, as we trailed in again fifteen minutes or more later. People waved, some snarled, I fear, and some ignored it all, toiling in lush green fields. We felt conspicuously and uncomfortably like royalty, passing down these empty roads, seeing the crowds kept back at gunpoint. After royalty like us had passed by, all returned to its usual mayhem behind us.

Luxor

What a great campsite; fortunately we did not get a puncture on the way while in the convoy. We parked in a shady garden behind the hotel. An ancient pink Hanomag truck had taken the best spot under a tree, its proud, single owner fed up with Egypt and heading north for Jordan and later India. In a corner under a smaller tree was an Irishman in a small tent with a bicycle, heading nowhere. A temporary resident of Luxor, he lives there for three months a year. And where were we going? 'Aswan, and maybe Sudan,' we said. At this point we really didn't think beyond the ferry; it might all end in tears at Aswan. We took in some sights, drank copious amounts of tea in the shadow of the Temple of Luxor and got generally irritated about rip-offs. I mean, you expect it, but not all the time; bread, tea, you name it, the price was always different. 'Welcome to Egypt, I give you good price!'

Temple of Luxor, Egypt

Another day, another convoy. At least this time it was possible to stop on the way. So we visited the magnificent temples and ruins of Idfu and Kom Ombo on the way. Idfu was magnificent, stunning, despite the hoards of day-trippers. As soon as we hit Aswan the armed escort melted away into the chaos. Where would we stay here? The perennial problem. No campsite, no hotel with safe parking. No lonely oasis, no shady delightful palmeries to hide in, just concrete and walled compounds. We had harboured a notion of camping in the lush gardens of the Old Cataract Hotel; well, it had been possible according to an ancient guidebook twenty years ago. Now, we could not even get into the hotel grounds without paying an entrance fee. Death on the Nile. Choking by mass tourism and choking by security concerns, the death knell of the independent traveller.

As luck would have it, we found the wonderful Hotel Sara, just along from the Old Cataract, 3km south along the river. What a view! The generous manager took pity on us and we lounged in luxury for two nights at barely a touch over £8 a night.

The pantomime season began in Aswan the next morning. Oh yes we can, oh no we can't. The ferry does go on Monday, but will there be enough cargo to attach the barge? Wait till Monday...

Monday morning bright and early we arrived at the high dam. Four hours later we were still at the high dam and full of nervous anticipation. We were dealing with Mohammed; they are all called Mohammed. Yes, there is enough cargo; no, there isn't enough cargo. There certainly aren't any other foreign devils with cars, so it's not looking too optimistic. We sit, and sit. How much are we willing to pay for this once-a-week mayhem? Can we consider coming back next week? What about driving back to Cairo? All those convoys... and what about going to Jordan? It's too cold and, anyway, we did that years ago and were paid to do it. It's going to be pay up, Wadi Halfa or bust, we fear. The game is played out all day.

The Sudanese cargo manager, Mahmoud, seems to be the one who pulls the strings, although he claims he is consulting with a higher authority in the building conveniently out of our sight. The first Mohammed seems to be on the sidelines, but his cut must surely be in the pie somewhere. More cargo arrives and it looks promising. It's all an illusion. It's all conducted with a great air of honourable pretence on both sides. We talk about future overland business for the ferry barges and they say they really are sorry that there isn't enough cargo and we have to pay more. Of course we are totally over a barrel, and pretend that it's really too much, we'll have to drive all the way back to London. What a shame. We really had hoped to visit the Sudan, such a friendly country.

Two o'clock and a deal is done. It's outrageous, but it's better than going back to Cairo in a convoy. Four o'clock and we are still waiting to be loaded. There is so much cargo there is barely room for the Land Rover. A barge is lashed to the side off the main ferry. It's crammed with potatoes, tomatoes, cans of cooking oils and some other stuff too. Five white western faces stare down at us inquiringly.

We are expected to drive the Land Rover up two wobbly planks on to the deck of this narrow barge without driving straight off the other side. The chassis almost touches the deck and the brakes mercifully stop us ploughing into the lake. But the Land Rover is on board and so are we a few minutes later. They take us proudly to our cabin. Well, maybe they weren't that proud, but just welcoming as usual. What a heap of sh... and this is the new ferry. The door is unlockable, 'no one on this ferry is a thief, we don't have them here.' Well, maybe not. The loos have already overflowed and are awash with 20cm of water, or whatever. In order to use this place, you must climb around the floor level on raised bulkheads and stand in the dark, hoping that your aim is straight and accurate. Judging by the state of the paintwork around the hole, the men are not very skilled.

Up on deck, yet more cargo is being loaded on to the main ferry. Those passengers paying deck class will hardly have space to sit, let alone lie down. Amongst the merry throng on deck are three young backpackers. A dying breed in many ways, are they. Tristan is on his first trip alone and heading to South Africa. He has just $200 in cash, plus his uncashable travellers' cheques. Christian is heading the same way, and a German, Johannes, is hoping to get to Cameroon and Chad via western Sudan. With a civil war going on in Darfur, his chances are not good.

In a cabin close to us are Camilla and Renaud. She is very English and he is very Swiss. They are planning to follow the Nile to Khartoum and then fly on to Tanzania to seek out the route followed

255

by Livingstone to Lake Tangyanika and then to Zambia. They are not so young. He is a sprightly 80 years old and she celebrated her 70th birthday in the Egyptian desert. Camilla, it turns out, was the former secretary to the disgraced politician, Profumo. It's a long time ago, but we too are old enough to recall the story. In fact Camilla left her job just before the scandal broke and, having met Renaud, moved to Switzerland.

As the sun set in true travellers' fashion, the ferry glided out on to the still waters of the lake to begin its 20-hr journey to the Sudan. The red ball of fiery sun faded quickly behind a rugged black line of rocky hills. The lights of the high dam melted into the waters and we headed out for the unknown frontiers of Sudan and Wadi Halfa. Nothing could spoil our excitement now – this was the point of no return!

No Going Back
There was something rather unnerving and final about this ferry trip. Up until Aswan, we could have abandoned ship and returned to Europe. But now, there was no going back. It had been nearly thirty years since Bob had been to the Sudan, with only good memories of the country. The hospitality had been amazing.

Despite the filthy conditions, we slept well and arose early to see the famous Abu Simbel monuments quite clearly across the water at dawn. Later, as the ferry crossed into Sudanese waters, a speedboat shot up to us. Was this the police or the Customs? We could only stare in disbelief as the cables tying the barge to the ferry were disconcertingly unleashed, and the Land Rover sailed off into the distance. We'd paid too much and now they were stealing the vehicle! We suspected that some cargo would disappear from the barge en route, but hopefully the old Land Rover would not appeal to anyone else.

Wadi Halfa
We land at a long jetty and disembark, following everyone else to the immigration but not much is happening. At least the building is new and they are selling hot sugary black tea in surprisingly clean glasses. It takes ages to get our passports stamped, but everyone is pleasant. Welcome to Sudan, have some tea on me.

A largish man who could be anyone's jolly old grandfather came over. 'Where is your car? Please get it now.' Kamal Hassan Osman is the Customs man, and he wastes no time. We quickly rush back to the ferry barge and somehow drive down two more equally wobbly planks to get the Land Rover on to the narrow jetty. The brakes still work. The Customs paperwork is quickly sorted out and then we wait. Some other work has to be done. We wait and wait. Now what's wrong? It's getting dark too. We sit and brew up more tea and more tea.

And then without explanation we are off. Oh yes, can we give our friendly Customs man a lift to town? Is there a choice? Off we go; we are not heading into town but over some low dunes to the next bay around the lake. Here is a small settlement set in a dusty sandy valley with some irrigation near the lake. We stop outside a walled mud house. It's not much to look at from the outside, but inside it's a treasure. There are exquisite bone china tea sets, cut glass, elegant wooden dressers and family photos. This is the house of Kamal Hassan Osman's sister. We take tea and biscuits, hardly daring to breathe in case we drop a china cup. He speaks English quite well and we are able to hold a decent conversation. They show us some photos of the old town of Wadi Halfa, before the Egyptians decided to build the dam and flood the Nile valley upstream forever. Wadi Halfa has never recovered from its flooding, and there is obvious distress about the loss of their old homes and the beautiful oasis now below the waters.

After tea, we took Kamal Hassan Osman to his own house in the sprawling desert town of new Wadi Halfa, doing the rounds of shops, to buy fresh bread and cheese, and friends en route. We were starving by now, and he invited us to have dinner at his house. His house was inside a large walled enclosure, with many rooms carpeted by sand. A big fridge took pride of place, with basic furniture and some decorations, posters and old photographs; otherwise it was a simple set up. Various nieces and nephews seemed to call by all evening and eventually we had a superb meal of beans, rice, spaghetti and a little goat. Copious amounts of tea kept us going as he enlightened us about things from his point of view. We were invited to a wedding party, but unfortunately it was rather late and, after our sleepless night on the ferry, we were rather out of energy. Our bed in the Land Rover outside was very welcome.

The next morning was spent changing money, and obtaining a tourist 'registration' stamp from immigration, which cost so much in extras to officials that we had to go and change more money. We met Camilla and Christian; everyone was having the same problem, so many offices to visit and so many bits of paper. Finally we were all set to leave; the others had to wait another day as the train to Khartoum had been postponed. In fact the whole town was shut down, as several weddings were taking place – there was no transport in or out of town that day! So the party we missed must have been a big one! Their hotel – the Nile – offered scant comforts and the train tickets were surprisingly expensive.

Our problem was different; where was the road to Khartoum? Maps are a wonderful thing until you realise that a road on a map in northern Sudan is in reality the sand beside the railway, and on which side of the line does it go? This road is a complete figment of Michelin's imagination. We followed the railway line but, even before leaving Wadi Halfa, we were in quite deep sand and then vehicle tracks led off to the north. More by luck and guesswork than by using the map, we found a way. The actual railway tracks were not visible because of the sand, but luckily for us a line of telegraph poles followed alongside the railway tracks. We tried to keep as close to the telegraph lines as possible. For 300km we followed the tracks; occasionally the sand was very deep. We met just one vehicle all day and that was only 15km out of town. Despite the fact that there is a railway here, which offered some comfort, this route is surprisingly remote and devoid of life. With a new road now linking Omdurman and Khartoum with Dongola on the Nile the trucks now head off from Wadi Halfa to Dongola, leaving our route disused.

At Station Four we encountered some workmen preparing the line for the expected train. They were sweeping sand off the tracks, and were not pleased with us as we pulled into the station astride the lines, carefully avoiding the points and steel cables. 'Which way to Khartoum?' They were not amused and directed us out into the deepest sand a hundred metres from the station. Almost immediately, we were horribly bogged in deep sand. At this point we discovered that the low ratio gears on the Land Rover would not engage. Fortunately some digging and then diff lock got us out. Well away from, and out of sight of, the station, we again headed as close to the railway tracks as possible, sneaking on to the embankment at the next deep sand. Sand, sand and more sand; staying close to the lines was often a recipe for a serious bogging, so sheepishly we took to the track itself. This was surprisingly smooth as the rails were buried by hard sand and the only problem was avoiding touching the lines themselves. It was lucky for us that the train comes only once week in each direction, as we were stuck on this embankment for 50km with no way off.

It was great to camp wild in the desert at last, no one in sight. After circumnavigating Station Six at a distance, we drove well off to the east below some low rocky hills with a tremendous view west for 80km. Having just parked and made a cup of tea, Bob said to me, 'Look, what's that over there? It looks almost like a train!' 'It is a train,' I replied. It was a large freight train, which obviously had not been invited to the wedding party!

For much of the next day we followed the lines again, and reached civilisation in the form of a town called Abu Hamed. We got completely lost here. The road into town disappeared into a rubbish heap and as for the road out, well that was anyone's guess. We drove over some railway lines and that at least woke someone up. The green lushness of the Nile was not far from the station, but fear of being stopped and asked for papers and the inevitable donation to the local constabulary persuaded us not to linger for a photo stop.

A smart Toyota drove past and into an equally smart building, a foreign aid food office. We stopped and asked the way. 'Follow me,' he replied kindly, and drove off again. In good English, he said, 'You go that way; see the white posts? They will show you the way.' Once more the 'road' was literally tyre tracks between posts every kilometre or so, but at least the cylindrical posts with triangles on top were clear on the horizon. Again the sand was surprisingly deep. We camped in some dramatic boulder fields, the most stunning camp yet. The rocks looked like ghostly turtles and gigantic beaked crows in the pale moonlight. The rocks all pointed the same direction, cut by millions of years of wind erosion. It was a wonderful place to spend the night, so lonely yet so peaceful. We switched on the BBC to hear reports of conflict in Darfur and could hardly believe it was the same country.

Our desert crossing was soon to end; we drove literally right along the side of the Nile. The verdant and lush irrigated fields beside the muddy sluggish river contrasted abruptly with the stark dark and harsh browns of the desert. We stopped for tea in the Land Rover and walked over to the riverbanks to watch a felucca sail by. The children and men working in the fields rushed over to say hello. Their outstretched hands merely wanted to shake ours. Such a change from Egypt. No bakshish was required or demanded, but we gave them some empty plastic water bottles, which are very valuable out here. Arriving in Berber, the first real town for hundreds of miles, we found fresh bread and diesel, perhaps two of the most desirable commodities we could have wished for! And from here on the tarmac began! We felt excitement at not having to bump along anymore, but great sadness at leaving the desert.

Desert camping in Sudan

Pyramids of Meroe, Sudan

At Atbara we found the new road to Khartoum, a toll road and a super highway supposedly paid for by Osama Bin Laden when he was staying in Sudan before moving to Afghanistan. This road melted away the miles and soon we arrived at the famous pyramids of Meroe.

Although much smaller than their famous cousins in Egypt, these dark-coloured tombs present an impressive sight at sunset, sitting on top of vivid orange sand. We camped a mile away on stony ground below some sheltering hills. Another day another puncture, just to spoil the tranquil evening. Some local camel herders came by for a while, hoping to sell old coins, stone carvings and all manner of decorative jewellery. There was none of the forced slick selling of the Egyptian merchants, just a quiet charm, still enough to do a little business though. And anyway we were camping in their desert backyard.

And so to Khartoum; the very name is enough to conjure up visions of the exotic. We camped at the Blue Nile Sailing Club, a large car park next to the Blue Nile. One redeeming asset was a shower that worked, and after three days in the desert that was truly essential! There was also a pleasant green lawn with shady trees, where we met some fellow travellers – Ina and Alex, who were planning to cycle across Africa over a period of two to three years. But they had already been on the road for over a year, and were about to fly back to Germany for Alex to sort out his tax returns. He was constantly doing something with his laptop, and we seriously wondered whether they would ever fly back out – to Addis Ababa. They must be crazy! Even crazier than us, we thought!

Then another white man arrived with just a tiny bag of personal possessions. He was Reinhardt, in the process of shipping his vehicle from Aqaba to Port Sudan to avoid getting a carnet for Egypt. He planned to drive his almost new Land Rover into the desert and make a film following the progress of a particular tribe of Sudanese nomads throughout the year. This would mean several trips to the country at different seasons, and he planned to ship the vehicle out and back each time. On the vehicle, at least when it left Aqaba, was all his expensive camera equipment, not to mention sleeping bags and other essentials for life in the desert! How much would remain when it arrived in Port Sudan? His girlfriend was due to fly out to Khartoum at the weekend, then they would take the weekly public bus to Port Sudan and wait for the ship to arrive. To get back to Europe, he was planning to ship the vehicle to Jeddah in Saudi Arabia, but, not being married to his girlfriend, she could not travel through Saudi with him, so would have to fly to Aqaba and wait for him there. The whole thing sounded far too much hassle – why not just pay for the carnet for Egypt and enjoy the time together?

> **Mirror, Mirror on the wall,**
>
> **Who is the craziest of them all,**
>
> **4X4 drivers, Motor bikers or cyclists?**

The main feature of the campsite, and marooned within the compound, was Kitchener's famous boat – the one that was carried literally from Cairo to Khartoum in bits to circumnavigate the various cataracts that blocked passage up the Nile in the late 1800s. Although the innards of the ship have long since gone, the basic hull, the gun and all decks still remain intact.

We spent three fairly restful days in the city, visiting the newly painted Mahdi's Tomb and Khalifa's House museum in Omdurman across the river. It was the Mahdi who laid siege to General Gordon and ultimately killed the 'Britisher'. It was equally fascinating just driving and walking through the city streets and markets, people-watching.

In the Grand Hotel, magnificently situated on the riverbank in Khartoum itself, we talked to a waiter from Juba in the south of the country. 'Oh, you've been to Juba, Sir? My home is there. I want to go home but I daren't go back yet. Too many people have no father, no mother, no brother or sister...' We couldn't afford a drink here, but it was a wonderful place to visit, and an unbelievable contrast to the poverty outside.

Having heard that Camilla and Renaud were meeting someone somewhere in Khartoum on 25 February, we went to the Acropole Hotel on the off chance they might have been seen here. Two quite old white foreigners might just be noticed in town! 'They're not staying here, but they are meeting someone here in half an hour,' said George the Greek, who has owned and run the hotel for forty or more years. George is a well-known 'institution' in the capital and is a mine of information. The Acropole is a very plain building on the outside, but inside it was a haven of tranquillity, and we were happy and comfortable while we waited. Most of the guests seem to be aid workers.

At 10:30am Camilla and Renaud walked in, amazed to see us again. We swapped stories, then the young man whom they were meeting, and who works with their daughter-in-law in Basel, Switzerland, insisted on taking us to lunch as well. It was not perhaps the sort of restaurant they were used to in Switzerland – a hot sweaty bean restaurant, crowded with hot bodies. With a low roof, crowds of clientele and with huge fans whirring, this place was like a steam bath, though the food was surprisingly good. Their journey had proved eventful; the train was fine, if a little slow, and they had hired a car to visit the ancient sights north of Khartoum.

Kitchener's boat in Khartoum

The friendly people of Sudan

It's quite easy to breeze by, cocooned in your own vehicle, passing villages. Passing people whose lives are so different, clinically isolated from them by the very thing that gives you the freedom to see these places and people. Wherever we repaired our tyres, someone would speak English and ask us many things; sometimes we were offered cold drinks. The Sudanese have retained their great sense of self-respect. In the meantime the media are making out that the country or its government is full of terrorists. What is the truth of the Darfur rebellions and humanitarian crisis? You will never know the story as a traveller. It is hard to imagine what atrocities have occurred, when all our experiences locally have been warm, positive and welcoming. The western region of Darfur hides the wild and rugged Jebel Marra a place that featured in an early Exodus brochure in 1977. A few overland travellers made it through to Chad in the early period of the new millennium, but conflict between farmers and nomadic herders turned into a desperate war. Will anyone ever be able to climb the peaks and explore the canyons of this remotest place in Sudan?

The Mahdi's Tomb, Omdurman

A super new highway runs south from Khartoum to Wad Medani through the cotton-producing belt. All this is irrigated from the Blue Nile to the east and the White Nile to the west. After Wad Medani the countryside is dotted with more traditional thatched hut villages and the low hills give way to vast swathes of tinder dry yellow grasslands. The hotel at Gedaref had no parking, so we camped out south of town in the tinder dry grass, hoping the distant bush fires were not spreading our way. Another flat tyre, but the road was no longer the narrow muddy track that Michael Palin had to follow some years ago on his Pole to Pole expedition. Now the gravel road was unlikely to get muddy, even in the rainy season. Another potential problem smoothed away.

The rusty roof of Africa

Suddenly we were at Gallabat, on the Sudanese border with Ethiopia. Neither of these two villages was more than a collection of thatched huts, with some new tin-roofed huts as well. Almost immediately a precocious boy came up and wanted to change money, then he wanted a lift, then he wanted more money and then he refused to give us even half the rate. As we entered Ethiopia we were asked to fill in a form saying how we had found the immigration staff – courteous, helpful etc. etc. … Since we had entered the country one day before our visa had been 'issued' in London, they were indeed very helpful.

The gravel road wasn't bad and the scenery became steadily more spectacular as we climbed up from the blisteringly hot grassy plains, through lightly wooded bush and low hills into sight of the great Ethiopian escarpment and plateau lands beyond. We found a great place to camp high above the road with a stunning vista ahead to the mountains. Unfortunately, despite the appearance of being miles from anywhere, two local woodcutters found us and sat down silently in front of us as we had dinner. Then the begging started. No words, just hands out. Their machetes suddenly seemed potentially menacing. Then some soldiers came by with large guns. It all started to look less secure than it had seemed. One soldier ran off into the bush, while the others continued up the track, then the woodcutters started begging again. That was it. No chance of having a quiet night here.

Reluctantly we drove on up the magnificent escarpment, passing spectacular drops and quaint villages on the way. People shouted at us rather alarmingly through the villages and some kids picked up stones, but no one actually threw one. This was a big change of atmosphere. 'You, you, you,' shouted the kids. Around 9pm we reached Gondar, being somewhat nervous by now and thinking why the hell were we doing this. Then some local youths chased us all around the town square as we sought to find the Hotel Terara where we could camp. It was not a good introduction to Ethiopia.

Gondar castle, Ethiopia

In the morning things looked a little brighter. A large overland truck with bald tyres and no clients was parked at the end of the hotel gardens, the first overlanders we had met. The clients were all off in the Simien mountains trekking while one driver and two sick passengers were staying in the hotel. The driver was Debs, a lively woman from Australia and more than a match for any African trickster.

Our fuel tank decided to spring a leak at this point, so a day was lost sticking Araldite glue all around the seams to prevent a drip becoming a torrent. We drove north on the road to Axum and soon decided to abandon any notion of doing that 650km diversion. The road was appalling, rocky and rough tyre-eating stuff. The famous and spectacular castle in Gondar had changed little since a visit in 1976 and there were, if anything, fewer tourists in town. We visited the incredible painted ceiling of the Debra Selassie church. By day, Gondar seemed a friendly enough place, but there was always the incessant begging. And most of the people were not really beggars. Having sorted out some insurance and bought a new tube, it was time to get back on the road.

The road to Bahir Dar was being improved but only 20km was actually completed, so it took us 6hrs or more to do the 200km trip, bouncing over rocks and rutted red mud. All along the way, people inhabited every square kilometre of the terrain.

We finally bought some more good tubes in Bahir Dar to ease the tyre crisis; the tank still leaked, but not much. The Hotel Ghion was a great place to camp with lush vegetation, beautiful gardens and shady trees overlooking the lake. It also had very cool pleasant rooms, which we treated ourselves to for the second night. A German couple were camped here; their truck sat with its springs in bits. They had been on the ferry a week ahead of us and it turned out there were ten other foreign vehicles on it.

Changes in Ethiopia
Ethiopia is really different, that's for sure. It's rather a sad place though; initiative seems lacking or stifled. There is hopelessness about the future and large Food Aid tents dot the countryside. That perhaps is one problem in itself, too many handouts and no self-help. The women seem to be doing more than is reasonable in the work stakes; the men just seem to sit and contemplate. Lake Tana is a vast lake and surely a vast reservoir of potential food-giving irrigation water. Yet you can hardly find any vegetables anywhere; no carrots, no tomatoes here. If you like chunks of garlic on their own, that's fine.

On a later visit in 2014, a vast change had taken place in the country. It seemed dynamic, go-getting and had one of the fastest growing economies of Africa. What had changed? Probably a lot less foreign interference and peace after the war with Eritrea.

The road to Lalibela started out being quite good, but barely 15km off the main road it deteriorated into the typical rock-rutted vehicle-crunching boneshaker. Only 25kph was possible on this road. At this speed the local village kids could run faster than us, especially going uphill. And we were always going uphill! The countryside was dry but not totally barren; trees and villages dotted the landscapes. The road followed the highest ground along the top of the plateau, and on both sides the land dropped precipitously into river gorges and wide-open canyons.

'You, you, you!'; it never stops. Luckily for us, one of the small villages along the way, Mulu, had an adequate hotel with a walled compound. It's hard to imagine having to sleep out on the roadside around here. We had yet another flat tyre to round off a knackering day, but plenty of small boys to clean the Land Rover. At least these boys wanted to do something to earn some money, and we were more than happy to oblige. The next day at the turnoff for Lalibela, we met two backpackers desperate for a lift, but we had no place. About an hour later we came upon the two Europeans, sitting under the skeleton of a thorny tree looking pretty desperate. They had been dumped by their lift and it was still 25km to Lalibela. So we piled their bags in the back and they rode on our roofbed platform. They were from the Czech Republic, and were backpacking around Ethiopia for three weeks.

At the end of the tarmac road, which leads only from the town to the airport, we camped in the driveway of the Seven Olives hotel. Bob had managed to sleep on a veranda here in 1976, having met the American tourists who had the room. We shared the driveway with a couple of Swiss vehicles, the first we'd really seen doing the overland.

Siân at Beit Giorgis, Lalibela

The Rock Churches of Ethiopia
Lalibela had grown since 1976 and most houses appeared to have electric wires attached. There were more tin roofs and less of the picturesque two-tiered round thatched-roof huts. Some of the churches had somewhat hideous but necessary weather-protecting roofs over them; otherwise not much had changed. Elsewhere in Ethiopia, other virtually inaccessible rock-cut churches are found in Tigrai Province near the town of Hawzien. The churches of Abuna Yemata and Maryam Korkor are two particularly dramatically and scarily located pilgrimage sites.

We climbed to the roof of Africa and then plunged down the eastern escarpment into the green and semi-tropical paradise called Woldiya. Two foreign aid workers sat nearby as we slumped down in sumptuous armchairs at the hotel and camp. They seemed to be solving all of Ethiopia's problems over a cold beer or two, such reassurance. Later they left in their latest model Toyota Land Cruiser with its radio aerial strung high over the bonnet and roof.

In the local market there were plenty of bright green water carriers and washing powder everywhere, but no food. What do they eat in this country? We eventually discovered that the hotel had its own bakery and huge vegetable gardens behind the building. Here was a veritable Garden of Eden with carrots, tomatoes, avocados, coffee, papayas... So why couldn't anyone except this enterprising manager of the hotel grow food? Again we had little or no sleep, all night a great wailing of Coptic prayer drowned out the faintly lethargic crickets and cicadas.

Tarmac at last. Yes, the map said so and it was, well for the first 50km and then abominable gravel once more into Desse. We stayed down the hill at Kombolcha in a wonderful hotel. We decided to splash out and have some local food in the café. Yes, the large sour flat moth-eaten looking cold pancake – Injera – and accompanying hot spicy beans, yellow beans and lentil-like beans.

The barren bottom of Africa
The road wound its way down and down, but it was not as spectacular a drop as we had expected; more a gradual rolling down and around into stifling hot winds and scrubby acacia bushlands. An occasional village seemingly grew out of the bush. The houses here were more like beehives, rounded mats woven together more like tunnels than huts. The people grazed thin, very thin, cows and goats on the tough spiny bushes and grasses. The people here are the famous Danakil or Afar tribes who terrorised the early explorers in the region, Burton and Thesiger. They terrorised us a bit, to be honest; some waved but most did not. Breaking down around here would not be a good idea. We followed the Awash River for a while and then decided to pull off the road about 100m up a track for tea and lunch. We had peanut butter and a tin of vegetable salad.

Halfway through our drinks, a plain white Toyota pulled up behind us. Five men got out, with large guns at the ready. That's it, we are going to be robbed, it's finally going to happen. We did our Salaams, and Good Mornings in case they weren't Muslims. Bob shook hands with them and I nodded my head respectfully from the interior of the vehicle. No one spoke English, but it was intimated that we should drink up and get back on the road immediately. Wow, was this scary? They could have smiled a bit. Well, we hadn't yet been robbed yet. 'Police' they claimed, but it was not in any way obvious. They were probably the local warlords.

With our adrenaline levels at an all-time high, we gulped down the remains of our tea and got on to the main road, where we stopped at the next settlement, a small scruffy collection of permanent but temporary-looking shacks called houses. At the pharmacy, the only proper shop in town, the man spoke English and translated. We were obviously international spies. Even spies have to stop for tea. It was all quite absurd, but no one demanded any money and, after further consultations, the episode was closed. Needless to say, we did not stop again on the side of the road until we reached a fuel station.

Qat
Qat is a plant often described as resembling an English country garden privet hedge. As the sun hits its strongest period, that is the time for the whole of the Horn of Africa and neighbouring Yemen to start the daily ritual of munching and chewing Qat. At least the fighting is probably halted for this brief period! Nothing can disturb this ritual; well, almost nothing. Exceptions are apparently made when a foreign tourist appears on the scene, such a rare occurrence is this.

Just before the border with Djibouti the road suddenly cascades down a fantastic volcanic wall, the remains of a long extinct caldera with dark brooding lava cliffs and a brilliant white and pink dried-out lake bed. This salt lake stretched into the dark, almost blue, mountains beyond. It was lunchtime at the border, even though it was now well past two o'clock. The Ethiopian immigration officers were very helpful, but customs were not to be found. Eventually a soldier sheepishly took us to one of many rusting cabins, old cargo containers and railway carriages. Here in air-conditioned comfort sat the elusive customs crew, cross-legged and chewing away to take their minds off the sweltering heat outside. Papers were duly signed and stamped in two minutes and off we went to the Djibouti border a few kilometres around the hill. 'Bienvenu à Djibouti.'

Things seemed much more organised here on this side of the Galafi border. The scenery continued to be outstandingly rugged, with distant black craggy ridges and vast plains. Along the road were beehive-shaped dwellings and a couple of scruffy settlements of makeshift materials. Along the road, the people seemed friendlier than their Ethiopian neighbours. Ethiopian trucks rattled along this well-made road to the port of Djibouti. With the border to the Eritrean port of Assab shut, Ethiopia is now totally dependent on Djibouti for its imported goods. Roadside cafés, brothels and basic makeshift facilities have appeared, to make life that bit more bearable for these friendly truckers. The heat in summer must be unbearable. Many of the trucks ambled along at no more than 70kph, some belching black smoke. Some sat forlornly broken down at the side of the road; others have rather worn tyres.

The first major town we reached was Dikhil and, to our great delight, there was an auberge with a large shady garden. Surely a good spot to camp but the price was pretty steep at $15. Well, we knew Djibouti was expensive for foreigners, probably because most are either military, aid workers or diplomats, and virtually all on huge allowances which inflate local costs. The hotel owner appeared later, completely stoned on qat, and with a cigarette drooping carelessly out of the corner of his mouth. He flatly refused to allow camping at such a price and then insisted we take dinner. Steak or steak or steak, and horribly expensive. His abrasive manner and glazed expression forced us to leave.

We continued on towards Djibouti town and then decided to try the road to Tadjoura, but by now it was 6.30pm and pitch black. As luck would have it, we finally found an amenable camping place on the beach at Ghoubbet. Some French military were just leaving, piled in the back of an old Bedford truck, having drunk the place dry. There was no water, but what the hell. It was a stiflingly hot night despite the sea breeze.

Salt crystals surround Lake Assal, Djibouti

In the morning, bleary-eyed and tetchy from the long drive, we headed for Lake Assal nearby. The road snaked over amazingly broken lava fields past currently dormant volcanic cones. The sheer number of old cones was breathtaking and somewhat sinister. Lake Assal, although only 15km from the sea, is below sea level and only separated from the sea by lava fields. The heat was already stifling by 8am. The brilliant blue lake is surrounded by the walls of an ancient cauldron, multicoloured but mostly dark and jagged. Around the lake is an almost fluorescent white collar of super brilliant crystalline deposits. The road descended into this vast cauldron of colour and seemingly dead wasteland.

Yet life is found here too. Small antelopes hop agilely across the rough landscape, somewhat tentatively watching the intruders. Distant flamingos collect around the shore. Down by the water's edge are amazingly shaped crystals and rare volcanic rocks, Halite being the strangest, with its egg-shaped hollowed-out rugby ball shapes and brown crystals locking into each other. Brilliant yellow sulphur crystals grow from the water's edge. This is not a place to walk around carelessly without a guide. Here is the birthplace of a new ocean, where Africa is being ripped apart along its famous Rift Valley. Here Somalia and most of Kenya will be cut adrift by the forces from below the earth of Djibouti.

Local Afar women, Djibouti

The road east to Tadjoura holds more surprises; massive dried-out whirlpools of solidified lava, where hardy scrub acacias are trying to stay alive, more ominous-looking cones and then a huge climb up away from the lake and the coast. The Land Rover can barely cope with the gradients and the gearbox sounds noisy. On the top of the plateau, the views south of the narrow Bay of Ghoubbet are stupendous. The waters are almost cut off from the Red Sea. Towards Tadjoura the greenery increases, and soon we can make out the Foret du Day, a small zone of intensely forested hills. A freak of nature, born from the sea mists and very infrequent rains.

Tadjoura had barely two streets, the main road was unmade, the houses whitewashed but hardly picturesque. Hardly any people were to be seen, and those that were along the mildly interesting seafront were not too welcoming. The only two places to stay seemed as sleepy as the town, so we left. Perhaps the Foret du Day could offer a shady restful retreat. However with the prospect of a thousand-foot climb somewhere ahead, we decided not to take any further risk and retreated to the main road. Another option was to stay in Randa, but there was nowhere to stay.

Suddenly a thunderous deafening, screeching roar broke the silence of the desert. We jumped out of our skins in fright. The French were playing soldiers across the desert, and their air force was terrifying everyone with their low-flying supersonic fighter jets. Djibouti, it seems, is the playground of the French military.

Arta, located on top of the highest hill with a stunning view over most of Djibouti, is the nerve centre of the military, but tourists, it would seem, are not welcome. Siân went in to ask if we could spend the night there. The boss, a kindly black man from Reunion, was most embarrassed, and apologised profusely that he could not help. While he was trying to work out what we could do, a pompous little white Frenchman in shorts, Major Somebody-or-Other, said, 'No way, we are not a tourist hostel.'

So we dropped down the hill as night was falling, and drove into the outskirts of Djibouti town; very uninviting with the squalor of refugee camps, cardboard housing and vast areas of very poor habitation. The truckers were camped out in compounds, but we were not about to join them. A lot of beer bottles littered the roadside and we were very apprehensive about our prospects of finding a hotel or camping spot with secure parking.

All the central hotels appeared devoid of parking areas, so we headed in desperation for the Sheraton. Perhaps we could camp in the guarded car park?

The atmospheric style of Djibouti city

Chilling out in Djibouti

The hotel manager, Peter, was a charming gentleman from Switzerland, and following Siân's negotiating skills, he graciously let us have an AC room with TV for $50. It was way above our budget, but we were talking Sheraton. The TV had BBC WORLD and every satellite station known to the world. The security guards were a friendly bunch and totally curious about our journey. They were very pleased with a modest tip and went back to their kiosk very happy. In the hotel garden there was a banquet in progress for the foreign guests; that is the French and German military and no one else. Ordinary people do not come to Djibouti, it seems, which is a pity, for the country is truly astonishing in all its facets.

Djibouti town, though, was a big surprise; the central districts were full of colourful markets, old and new mosques and the old French quarter was a delightful collection of streets. The bay was pleasant and the eastern area was quite orderly, with shady streets. The old colonial quarter of town has some picturesque buildings, many with arched facades that give much-needed shade to vendors and pedestrians. There are street cafés and ancient tamarisk trees. Well-dressed businessmen mingle with entrepreneurial ladies dressed in amazingly colourful attire, laughing and gossiping over strong coffee. The African quarter and market area is totally absorbing. Shopkeepers and street vendors promote their wares, groups of men sit beside old whitewashed mosques smoking hookah pipes and cigarettes. The women in the clothes market are deep in animated conversation. We are probably the source of much of their humorous laughter and gossip, but no one is threatening. We have tea; it's the milky Indian-style tea, which is a big surprise, as there aren't many Indians about. There are lots of Somalis, Ethiopian truckers, Eritrean refugees, Afar and Danakil tribals, Yemenis and Arab traders, but hardly any white faces.

Check-out time in Paradise Hotel is one, but two is fine. We savour the cool luxury, watching the TV. Our world is one of make-believe; cocooned in this small space where we can travel in time and space. We are transported back to little England. The budget is being discussed – we don't know if it's over or due next week; the weather forecast is of course for rain and the cars are just as expensive as ever on Top Gear. Down in the car park the Land Rover is cooking, the steering wheel is too hot to touch and the guards are half-asleep in the great fog of humidity.

We head out of town, topping up the fuel tank as we go – we don't want to stop in any dubious place in Ethiopia – and depart for the border. As we cruise through Dikhil, the gate man at the hotel waves with a huge smile on his face. The owner is not to be seen; it's well past Qat time anyway. The border people are friendly and helpful, allowing us to camp close to the customs shack. It's still blisteringly hot and humid, and sleep is hard to come by, but at least it's secure.

We chatted briefly to some of the immigration staff on the Ethiopian side. They apparently do one-year stints at this God-forsaken place as part of their duties, living in disused railway carriages or shacks made of incomplete planks of wood. After the border we hardly dared to stop until we found a café with parking halfway to Addis Ababa. This hotel/brothel/café was a delightful stop. The ladies of the house were very welcoming and very amused with their light-skinned guests.

By mid-afternoon we had entered a more fertile region but the gearbox was still grinding. Sometimes we couldn't tell if the subtle noise change was real or imaginary. From Awash the road climbed up and up, passing rocky outcrops and lakes, with the acacia trees becoming more plentiful. Then we were into pastoral scenes and gumtrees, eucalyptus swaying in the breeze.

Nazret was a pleasant haven; the markets were crowded, the road was crowded and suddenly we seemed to be in an Ethiopia of much more vibrant character than the one we had left at Kombolcha in the north. The hotel gardens were well kept and the room comfortable. Most importantly, we felt an immense sense of relief to be back in civilisation.

The city of Addis was pretty welcoming too. It seemed to have expanded in all directions since 1975 when Bob visited last. The next morning was spent finding the central Customs, though with some apprehension. We had not been issued with the green papers at the Galafi border; the man with the key to the paper cupboard was simply not 'in station'. Here in Addis no one seemed keen to rectify this, and all from the boss down insisted the border at Moyale could sort it out. Then, to leave the compound, we needed another piece of paper. The gate men seemed to think the Land

Rover was new and had somehow avoided Customs duty, being in this Customs zone. 'How did this car get in here?' exclaimed the big man. 'We drove it in here an hour ago.' 'No, no, you cannot have done, where is your gate pass?' It was finally sorted out at the mention of the big boss and the carnet showing our entry at Galafi.

Time in Africa just drifts by. One spends so much time over bureaucracy, visas, mechanical needs, finding food, eating it and trying to sleep in noisy brothel gardens or in scungy cockroach-infested rooms or, if lucky, a shady garden behind a hotel, mission or car park. There is not much time to spare for just being a tourist.

In Addis we did at least visit a couple of interesting museums. One housed the famous Lucy, one of the earliest sets of human bones yet found. Very few tourists seemed to be in Addis, the one exception being at the Taitu Hotel, an old Italian colonial-style retreat where lunch was the event of the day. Here we found some tourists and one overland party in the car park area behind the grand façade of the place. The truck looked a bit the worse for wear, with prominent oil leaks here and there. The springs had broken and the number of passengers was not enough to push the truck out of any mud or sand. Overlanding as a business, it would seem, is well in the doldrums these days. Terrorists and insurance companies have seen to that.

South of Addis a good road drops down from the highlands into the Great Rift Valley of East Africa. We passed a lone Japanese cyclist in this road – we will never know his story. The lakes of the Rift are famous for their bird life and pleasant scenes. From Awasa the road climbed back up into a region of very steep luxuriant hills, heavily populated and bursting with tropical fruits. Banana trees and thick foliage encroached on to the potholed road. Some sections were very slow going; the holes were deep and the rebellious Land Rover gearbox seemed to be trying to tell us something.

Despite the mass of people in this region, we felt uncomfortable. The villages were poor, the towns scruffy, noisy and boisterous, the people not always apparently overjoyed to see us. After Dila the road was again bad all the way to Arge Maryam, another half-made small town. Once south of the town the scenery became more pastoral as the road dropped down from these almost isolated zones of tropical vegetation. The bananas gave way to thorn trees, the villages replaced by semi-nomadic herding settlements. Tall termite mud towers appeared and the soil became amazingly red. The herders with camels, goats and very long-horned cattle seemed more friendly, and waved with an air of confidence.

Yabello Crossroads Motel
Marked on the Michelin map 650km south of Addis is the town of Yabello. The road is better and there is a fuel station at the junction on the main road. This is one of the last fuel stops before Kenya. There is even a hotel with electric light, hot showers, a nice garden with a pet dik-dik baby deer, and a restaurant. Quite who needed all these facilities was not immediately obvious as we cruised into the fuel station. But a few minutes later, on attempting to leave, a ghastly noise – a heart-rending shriek, more like – rang out from the gearbox and getting gears proved haphazard. Our mood sank to rock bottom in an instant. The long hours of apprehension were now over, with a suddenness so swift and absolute. That grinding noise – it wasn't so imaginary, after all.

With shaking hands and a pretence of trying to keep calm, we sat at the hotel café over tea, contemplating our misfortune. At least we had had some good luck being at a comfortable hotel. Many miles further north or south would not have been much to celebrate. Not far to the south and east are the tribal lands of the Borena nomadic people who have an intriguing method of cattle watering and depend on wells of a very unusual nature. It seems quite a number of small tour groups are discovering these nomadic herders' cultural uniqueness and the hotel hosts these tour parties in some degree of comfort.

Around sunset the mechanic arrived and confidently expressed the possibilities of repairs. Sitting at a white linen-covered table in a corner of the terrace of the restaurant were two white men, having appeared just minutes earlier in a super white gleaming Land Cruiser. A vehicle produced in South Africa, and patently not having any mechanical niggles. The men's accents could hardly been more British if they had been actors. Now this was quite a boost to morale. Ben and Gavin were working on an aid project to preserve and develop the native trees and forests of the region for future generations of Yabello. The trees in question were the pristine-looking junipers found on the higher ground above the bush, which have wispy lichen hanging down from their branches in a Tolkien style. These trees have survived, despite the heat, and it is to be hoped that the plantations of these and other valuable woodlands can be saved from the ongoing desperate needs of the increasing rural population surrounding the town. A beer was definitely needed here to soothe our shattered nerves. Ben and Gavin were staying for the whole week and helping to keep the hotel running.

The Yabello motel, Southern Ethiopia

The mechanic, a cheery fellow, arrived and within three hours, with the help of Masale from the garage and a long pole, the gearbox was soon on the sand beside the Land Rover. A small bearing no more than 5cm across had disintegrated; it had literally ground itself away. Naturally, despite all the many spare parts we carried, this bearing was not one of them! Perhaps one could be found in Yabello? No, it could not, after a couple of hours. Maybe in Mega, the next village to the south, or surely in Moyale, the border town with Kenya?

Ben and Gavin turned up in the evening for dinner and we had another beer. It turned out we had a mutual friend, Bart Jordans, now living in Paro, Bhutan. Well, one would expect this, after all Yabello is the centre of the world between Addis and Nairobi!

The mechanic went off on the once-a-day bus to Moyale, hoping to visit some mates at the same time. Things were tolerably okay. We had food, water, a bed, some mosquitoes, but not too many, and a friendly manager and staff.

Full of hopes, we decided to go into Yabello to see the town, the bank and whatever else was on offer. After a long hot walk, a rackety old Toyota people carrier came by. 'Jump up, mister!' This people carrier, though, had hard wooden benches, open windows, a torn canvas roof and no springs worth mentioning. Still the locals were friendly; women off to the market, a boy late for school, and an old man about as stiff as we were by the time we arrived in the dusty main street of the town. Town was one longish street of low shacks, new concrete, tin-roofed shops, a few shady trees and a few men languishing in the rising heat. There were a lot of bars, considering the volume of potential clientele.

The mechanic had not appeared from Moyale but no matter; neither had any bus, truck or minivan come from the south either. Back on the veranda at sunset Ben and Gavin reappeared. Ben now lives in Addis virtually full time with his Ethiopian wife. She was currently in America with relatives.

He works for two charity aid projects and seems to enjoy life in the capital. These projects were due to run for a couple more years yet, and a lot more training of local people to take over was still necessary. Gavin was due to fly back to England within the week.

The mechanic duly arrived, but empty-handed, from Moyale. This was not the news we had hoped for. What next? The manager of the hotel informed us that we had to vacate our room as some important VIPs had arrived. We were bumped along the veranda to a large room, but this one had no inside toilet and the outside version was a very uninviting hole.

A bus leaves Yabello every morning sometime around dawn for Dila. Just as the sun began to glimmer on the horizon, the manager appeared – 'You'll have to get this public bus.' A few seconds later, the bus arrived, coming from the village up the road. The only seats left were, of course, along the back bench. The crowd were not happy, for the road to Dila is not a super smooth highway, the seats have no leg room, the windows rattle, but the driver – well, he was very good. So that was fine. We just had to sit it out without getting gangrene in our motionless legs and numb bums.

From being isolated and cocooned in our own vehicle, we were now thrown into the thick of African life. At least we could feel some sense of community with the passengers, all of us suffering the obvious discomforts of the bus. Being this close to the local people, we could observe the little things that make travel such a fascinating experience – even under the most arduous conditions. And these people do this trip many times, whether visiting relatives in downtown Dila or taking produce to the market towns beyond to Addis. Who were we to complain?

Six hours and thirteen minutes from the Yabello crossroads motel, we arrived in Dila. The driver had managed to tell us in sign language where we might find motor parts and dropped at the door of a small store full of parts. This little shop must have sold every conceivable size of motor bearing. For just £5 we found the bearing. We bought two for good measure. This all happened within ten minutes of our bone-wrenching arrival in Dila. Tea was now the order of the day, a joyous relief. Within the hour we found ourselves sweating madly with fourteen other hot bodies in a Japanese minivan built for seven bumping back south. But now, with parts in our pocket, we didn't care how rough or how hot it got. We were heading south.

This minibus terminated in Arge Maryam, but despite pessimistic protestations at the minivan stop, we felt our good luck would get us onwards again. Another minibus turned up and, with five of us already aboard with confirmed seats, they drove slowly around the town touting for more passengers. Slowly more appeared; some got off and were not seen again. Did these people know where they were going or why? Not long before dusk the rush for places increased and we departed. Two miles out of town and a puncture. The spare was no better, but on we went again. It seemed we had no spare tyre now, and the countryside, as we knew, was pretty devoid of habitation. This could be a long trip if any nasty thorns punctured the terrible tyres.

In the cool air of dusk we still felt exhilarated. The sunset was magnificent. Siân looked briefly out of the window and saw a huge herd of zebra, just wandering through the bush. So close, and this was not a safari, just a local bus. Two colourfully dressed tribal ladies in front of us had been on the first bus from Dila and we certainly kept them entertained on route. Quite what was so amusing about us, we never actually knew.

After thirteen hours of buses and minibuses we were dumped at the motel as the minibus headed into Yabello. Ben and Gavin were there to greet us, surprised that Dila could provide the parts but pleased all the same. The day had actually proved quite entertaining.

The hotel was now bursting with activity and full of tourists, so we slept in the back of the Land Rover on the garage forecourt. 'But pay me for camping here before you go to bed,' said our friendly manager. The group from Germany were to visit the Borena wells and also visit an isolated herders' village to the south. In the morning they had breakfast in the garden and soon departed in their gleaming white Toyotas.

The mechanic arrived soon, and within four hours we had the whole thing back together. We loaded all our gear and paid the mechanic and his mate at the fuel station. The bill for labour was £33, including generous tips all round. Without a moment to test the vehicle, we departed. Nothing else could be done with the gearbox anyway until Nairobi. It was surely not totally sound after such a lot of grinding of metal. And the ground-up bits, they had been flushed out with the oil and a magnet. Great swathes of furry metal, like a Christmas tree, appeared on the magnet. Too much to contemplate; what other damage had been sustained?

The scenery was dramatic; again the evening light captured the spirit of the magnificent vistas of peaceful bush, forested hills and herders' camps. The termite chimneys became white instead of red, and the villages more isolated and more picturesque. At sunset, 200km south of Yabello, we reached the border town of Moyale. We hurriedly left Ethiopia before they had time to think about the lack of green papers. It was closing time at the border; it didn't matter about the green paper anyway. Across the bridge and another country.

Welcome to Kenya
Being in Kenya in itself was exciting, but of course we were now about to cross the northern desert. This was the one section of journey said to be unsafe, rife with gun-toting bandits, unfriendly nomads and chronically bad roads. At the checkpost we were advised to take an armed guard for the desert crossing to Isiolo. Stories about this section were rife amongst the few travellers we had spoken to in Addis. A couple on motorbikes had been robbed of all their money while riding on the top of a local truck. Another traveller had been shot in the arm. Various local trucks had been hijacked and robbed. The truck/bus in Moyale was set to take a sandy route via Wajir to Nairobi. Was this a safer bet? The route goes horribly close to the Somali border and, of course, all the bandits in northern Kenya were foreigners – from Somalia, not Kenya.

Our armed guard or 'guide' wanted to go to Isiolo, so that's the way we went. We headed west, parallel to the border, south of some high hills that form the border. Dramatic peaks on the western horizon dominated the drive to a mid-morning village tea stop. After this village the road deteriorated into a ghastly rock-strewn track barely wide enough for one vehicle. A large herd of wild ostrich crossed the road in front of us at a stately pace. At less than 25kph, this was very slow motoring. For hours we could see the distant hills around Marsabit, but for hours we got no closer.

The island of hills in which Marsabit hides are surrounded by endless plains of desert and scrub in all directions. Alas, the campsite was a waterless empty field, down a steep hill, far from the security of the gate facilities and deserted of tourists. Then the fuel tank started leaking again from the bumps of the shocking road. The shock absorber bushes had totally disintegrated in just one day, then a troupe of baboons invaded the roadside we had just parked in. Our guard decided to stay in town, so he hitched into town with the park warden.

The tank was re-glued with the last metal glue and then, after a hurried dinner in the dark, we met the camp boss. Further shocks were yet to come – it would cost us twenty-four dollars to camp here, with no showers and grim toilets, so we left and drove down into town.

Jey Jey Centre is paradise. 'Do you have any vacant rooms please?' 'Yes, Madam, a single with bath is 500 shillings and so is a double without bath.' What more could we ask for? The shower worked, the water was hot, the bed was clean and not soggy and the lights worked too. True paradise. But upstairs in the restaurant, a real treat awaited. Although it was now way past our bedtime, 9pm at least, we needed a drink to relax after a somewhat hectic day. Smells of Indian-style toasted sweet bread and fried eggs cooked in too much oil wafted through the air. 'Can we see the menu, please?' A snack to savour, fried eggs on toast that could have been cooked by the super chef of the old Khetri House hotel in Jaipur or the Dak Bungalow in Varanasi. This was Indian breakfast food at its best. So how could we resist a repeat dose for breakfast?

A foggy mist enveloped Marsabit in the morning. The forest here seems to live on dripping heavy morning fog. The fog cleared to reveal an endless vista of green bush and scrub dotted with dramatic soaring granite lumps of mountain. All morning we crawled along carefully, nursing the suspension and the gearbox.

Amazingly decorated tribal people were walking along the road to the market. These tall, feline-looking people are the Rendille. Related to the Samburu and Masai, their fearsome reputation precedes them. They carry spears and wear silver and red-coloured decorations similar to the Masai.

Rendille children, Kenya

The road improved at times and progress was good. Archers Post has no tourists now. After Isiolo, the road was not much better, but we began a steady climb with Mount Kenya now just visible to the south-west. The snows peeked out of hazy clouds. The air was cooler, the countryside farmed and the landscape looked more and more European as we ascended. Some white farmers drove by and we began to catch up with a smoking English Army Bedford truck. Ahead another truck had broken down and was being carried uphill towards Nanyuki on the back of another truck. We actually managed to overtake the one smoky truck and then the two others. We couldn't resist a wave and smirk at the sight of the British Army ignominiously crawling up the hill, a sick vehicle more like a local vehicle.

Our plan was now to find a place to camp, probably in Nanyuki, but it was still quite early so we continued to Naro Moro, the gate for Mount Kenya. With cloud on the mountain again, the temptation was to press on to the Kenyan capital. We somehow felt that, once in Nairobi, all our worries would be over.

The distant hills and mountains west of Nyeri were cloaked in the heavy grey banks of clouds that mark the beginning of the East African wet season. The road was surprisingly rough and not well maintained; Kenya is having budgetary problems. And then the clutch began to feel lumpy and unhappy. With only 20km to go, we decided, perhaps unwisely, to continue, but by now there was no chance to camp or find a cheap hotel.

As dusk approached rapidly, the clutch was not getting any better and, just past the crest of a hill, the traffic ground to a halt. The clutch chose now to fall apart; the pedal brought no response and this in an area not far from the outer slums of Nairobi. Stuck in the centre lane in the almost darkness, we somehow managed to push the errant vehicle across lanes of traffic oblivious to our plight.

To our eternal gratitude and amazement, a smart young local couple appeared out of the melée of lights, horns, loud exhausts and shouting shopkeepers, and said this was not a safe place to stop, and what were we doing there. 'We have broken down,' we said, 'and we cannot move.' Charles and his wife, Martha, suggested we push the vehicle into the large Uchumi supermarket down the road with 24-hour security guards. They kindly agreed. Such helpful people you couldn't easily find anywhere, and this was close to the poverty-stricken ring of Nairobi's worst deprived areas.

Charles and Martha invited us to stay in their modest but comfortable flat in a compound a couple of hundred metres from the shop. This was more than just charity and kindness; we were truly touched by their overwhelming hospitality. They cooked on bottled gas rings and the shower was a very large bucket into which a large coil heater was placed by hand. Martha cooked a super meal, traditional fare including spinach and bananas. Their shy but curious five-year-old son, Titus, was a bit overwhelmed by these strange white foreigners staying in his house. Martha's teenage sister Angella was eager to ask some pointed questions. After all, why would any sane person want to drive from London to Kenya when the plane takes only eight hours?

Upper Hill
Upper Hill campsite is a pleasant garden and old colonial house up the hill from Uhuru Road, the main thoroughfare. Breakfast was our favourite meal of the day; loads of mouth-watering bacon and eggs, toast, butter and marmalade. Not many customers were camping and an old overland truck stood in the corner waiting patiently for a buyer. It could be some time, there aren't many overlanders in Nairobi these days. Joyce had been running this place for some years now, and it was a bit of colonial England on the edge of the rapidly modernising city. Some new skyscrapers have grown from the hills close to the Pan Africa hotel, Haile Selassie Avenue area. Insurance companies, oil companies and now the EU has its flashy new centre just down the road.

As the big high-rise developers moved to this hilltop location with a great view of Nairobi, Upper Hill hostel moved to a new location in Othaya Road, Lavington.

Next morning a small pickup truck was to take us the last 12km into Nairobi, with the Land Rover strapped to the rear platform. We thought the AA breakdown vehicle was going to break down on the way, it was so overloaded and slow up the hills! They dumped us at the main Land Rover dealer in south Nairobi. It was all very ignominious for a veteran overlander, but it was easier than trying to fix it on the side of the road in a dubious corner of the city. It was not a big job. The gearbox should have been done too really, but it was all horribly expensive, so we decided to press on with our Ethiopian botch-up for now.

The downtown area was said to have become safer recently, but taking a trip there after dark was not recommended. We needed to check out getting a Mozambique visa here. It was not a problem. In the former overland years Mozambique was on the list of 'keep out' countries. Visa were not issued; this was during the civil war and before the liberalisation of the nineties. We scurried about the city all morning and relaxed at the Thorn Tree Café in the New Stanley Hotel. Everyone had always done this on previous overland trips, and this trip was no exception. But it was surprisingly quiet at the café; just some South Africans, safari people or mercenaries in days gone by, perhaps.

The Land Rover was already fixed by mid-afternoon, so we drove it back to Upper Hill and celebrated with two large steaks at the hotel. Another Land Rover had appeared; a Dutch couple with South African plates. Arnout and Saskia had driven up from the south over the course of the last six months, taking in most of the major places en route from Cape Town. They had lots of time, which we did not. An email from the shipping company P&O told us that a ship was due to leave Durban for Southampton on 10 May; the next would be late May and that was too late for us. This was rather earlier than desired. We had to be at work in the Alps in June.

We decided to spend a couple more days in this tranquil haven. Our families back in England had spoken to us on the telephone and were obviously relieved that we had arrived in Kenya in one piece: the Danakil had not eaten us alive and the terrorist associates of Bin Laden had not snatched us in Sudan, the Ethiopian shifta bandits had not stolen our disintegrating vehicle and the Samburu tribals had not thrown spears in our direction.

Where is Kenya going these days?

The main roads have crumbled; new roads have not yet been built, but have no doubt been paid for. A new president is coming to grips with corruption. Nairobi has grown upwards with some very slick new buildings. The old Florida Nightclub, where Whitney Houston's songs used to blast out, has been replaced by a glossy new Bank and Insurance building. The population has blown itself sky-high and the shanties have multiplied faster than the bugs in the festering sewage canals of the north-eastern suburbs. The glossy banks are not serving the people well. The jewel of East Africa is a slightly sadder place these days. There are some successes though, as always, amongst the abject poverty. Food production is up and tourism should get back on track when foreign governments lift their restrictions on travel to the country. Its coastal resorts are suffering since the terrorist attack that shook the country.

The above was written in 2004, but today many things have radically improved in Kenya, except for the troubles inflicted by the Somali problem. As before – positives and negatives – is it ever a simple picture?

Masai tribesmen at Namanga, Tanzania

Making Tracks South

With the shortage of time and impending rains, we were soon across the border into Tanzania. Kilimanjaro was hidden in a swathe of heavy cloud with a dark, thundery squall to the west. At Ngorongoro the crater rim and most of the slopes were shrouded in thick cloud. As we headed back to Arusha, we were amazed to see the red Swiss Land Rover of Natasha and Mickey, who had been visiting the crater. We had heard about their adventures on the way to Nairobi; they too had been apprehensive about northern Kenya, but in fact nothing had happened on the way. They had more time and planned to take two more months driving to Cape Town through Zambia and Namibia.

Intermittent rains and hot sun punctuated our run south through Tanzania. The scenery was spectacular in the watery sunlight, green and blue of forest and mountain ridges. The views across the uninhabited bush were vast. At Morogoro it was time to stop for the night. We spotted a white engineer in the road-building team and asked him if there was any campsite in town. 'I don't know of any campsite, but you might try the Acropole. That's where most expats stay.' So we drove there and asked the manager if we could stay in the car park. 'You'd better ask the boss,' he said, and phoned her. A Canadian woman, she said she would be in later, and of course we could stay. The staff offered us the use of a wonderful luxury suite for showers and toilet facilities, so we felt completely refreshed with clean clothes and squeaky-clean hair that evening! As expected, the place was full of whites in the evening; mainly engineers, teachers and aid workers.

Just south of Iringa is Kisolanza Farm. This haven of peace sits at some 2100m. As we drove in, two young white women came across and said in perfect Sussex accents, 'Do come and have a cup of tea when you're ready!' The campground was very well organised and the steaks were tremendous. For less than a pound we gorged on best fillet steak, so tender it could be cut with a butter knife. A party of four South African vehicles and their ageing occupants shared the camp.

The evening was spent exchanging stories. Where did you break down and who ripped you off? How is your vehicle going? What sort of engine do you have? Jasper and Emma, Justin and Becki had met for the first time in Nairobi after completely separate journeys from England, but lived only five miles apart in Kent.

The farmhouse stands on a hill not far from the camp; the roof is thatched, the walls of local stone. Such a magnificent house could easily fit into a Dorset village. The farm had become rather rundown during the excessive socialist years of Julius Nyere, when landed gentry were considered undesirable in a left-leaning Utopian regime. New but admirably set bungalows have now been added to encourage eco-tourism, the new byword in development in modern-day Tanzania. Things change. Tanzania has, if anything, taken over where Kenya was going before the stagnation that marked the middle and later years of the Moi presidency. Kenya sadly seems to be heading along the plains, whereas Tanzania seems about to climb the lower slopes of the upper hill.

South of Mbeya the rain was torrential, the countryside steep and luxuriant. At the border, which was now a few thousand feet lower, the temperature was hot and the air muggy.

Lake Malawi is a very picturesque lake. Ringed by high mountains in the north particularly, the lake is a magnet for backpackers wanting to relax after the rigours of African bus travel. The camp at Karonga, the first main town in the north, was not so wonderful; the showers barely worked and the place swarmed with mosquitoes at dusk. We decided to try sleeping on our roof, as the back was plainly just too stuffy in this humid place. The tent was erected, the mosquito net installed. But the rains arrived in the early hours of the night. A gale blew and the rain poured. We ended up soaked in half an hour and the mattresses bulged with torrents of rainwater.

School's out in Malawi

In Sega Bay north of Blantyre is a pleasant new campground with some simple rooms. Cool Runnings campsite was run by a widowed woman exiled from Zimbabwe. In the bar after dinner we met a group of three young backpackers from Britain with a tough-looking South African man over for the weekend from Lilongwe. He worked for Avis, seeking out stolen Avis cars across the continent. If you hire a car in South Africa, the chances are one day when it's not so old it will end up reincarnated as some rich African's private car, thousands of miles north of its original home. Previously he had worked with the South African police force; now he sits in casino or five star hotel car parks, for example, and looks for the secret markings that identify an Avis car. Then of course he has the unenviable task of persuading the local police that the car is stolen. If the owner says they bought it in good faith, he can do nothing.

It was a wet and soggy day when we rolled into Blantyre. Doogles, the main backpacker hangout in Blantyre, is a bar behind the main bus station. It is a pleasant town, its cool climate a product of the hilly regions it sits on. The rains were still rumbling around and the views of distant Zomba and Mt Malange were obscured.

We met up with Dr Liz Molyneux, a doctor associate of Siân's sister, for a cup of tea poured from a well-insulated teapot. Its cosy looked more like a hat worn by some Rastafarian characters we had seen in Ethiopia. Arranging this encounter took some imaginative work on the part of the longhaired barman manager. 'My cellphone has no credit, but if I call her twice and ring off before she answers, she should get my number and call back.' Sure enough, she did just that.

The road from Blantyre is good as far as the border; the scenery is again green and lush with massive rocky outcrops offering vantage points on route. The border official tells us our insurance is not valid in Mozambique. The yellow card clearly says it is valid for Mozambique. We are clearly not going to get across this border with the Kenyan insurance certificate. All the guidebooks remark on the corrupt nature of the road checkpoints. We had better shut up and pay up before they refuse to welcome us.

Great scenery going into Mozambique

Tete sits on the wide Zambezi River. The mosquitoes are out and the restaurant at the hotel we have managed to camp behind is stinking hot. The food isn't so great, but the Indian or Pakistani family who run it are helpful. They have never been to the subcontinent, which is a surprise. Two turbaned gentlemen are having dinner, but no other guests are seen.

According to the Michelin map, we can get to Chomoio in one day; a few towns are marked along the way. But where are these towns? We are in remote scrub and savannah forest all day. Occasional rocky outcrops and isolated villages appear, but the towns are deserted or smaller than expected. Mozambique is proving to be almost more remote than the Sahara. Towns are abandoned, fuel is rare, people are living a subsistence lifestyle. Very few vehicles use the road – a truck or two all day. We are quite shocked at the remoteness of it all, thinking we'd done that already before Kenya.

Chomoio was a strange incarnation. We stopped for tea in an out-of-town café, thinking we might camp there. It was full of heavy well-built men; whites enjoying some hearty steaks. They looked like ex-farmers from Zimbabwe and Portuguese expatriates still living in the country. Or were they mercenaries? Discretion being the better part of valour, we decided to move on! In town we sat camped behind a strange hotel where a windmill had been converted into tacky rooms. The windmill was not real but the café offered cold beers and a couple of well-dressed backpacker tourists. Things seemed to work in this town. It has become a refuge for many whites from nearby Zimbabwe; that is clear.

No one travels much in Mozambique. Fuel is expensive, there are almost no private cars except the big 4X4s of the white expats. Buses are rare, just one or two a day between major towns. Again we were on a lonely road and this was the main north–south road. Most of the beautiful colonial houses were abandoned, or their verandas used as flat areas for open market stalls.

We arrived in Vilanculos expecting some ritzy beachfront developments, flashy hotels and hundreds of beach trippers from the south. Plainly that development has not materialised yet. The road down to the main beach is a sandy track and the hotels, where they exist, are small and not-at-all stately. 'Town' is hardly much more than a couple of fuel stations, a small shop and little else. The big campsite is very well arranged, though. A big white South African man runs the place and all his customers are big white South Africans with their families, wives, lovers and expensive rubber dinghies. Most are heading out across the Indian Ocean to the nearby Bazuarti islands for diving. This ocean of material wealth and brash sea of glitzy 4X4s sits in a near desolate landscape of subsistence. The economic apartheid as clear and obvious as the one dissolved ten years ago in the south.

Three women walk by, carrying wood for their evening cooking. The beach is deserted, for it's quite rocky. Some African boys come by selling freshly caught fish. We successfully sleep on the roof again, and contemplate the strange divide of which we are part here in Mozambique.

Further south is the town and former Portuguese settlement of Inhambane. On the way we drive through lush and extensive plantations of palm trees. Inhambane lies some way off the main road on a peninsula renowned for its beaches and laid-back lifestyle. Laid-back it certainly is; the beaches are good, the surf exciting. The camp areas and beach huts are quite basic, though. The better resorts are hidden away with exclusive beaches. Outside these compounds is another life.

Inhambane is a very strange place, deserted in part, regrowing in part. The church square is deserted, the streets like a French country town on a Sunday afternoon without the cars parked outside the bungalows. The mosque is old but rather tatty. The locals are all to be found outside town in half-built shanty areas, with colourful markets ongoing all day. The locals plainly do not have any ambition to live in the clinical and empty town of the former colonists.

The night on the beachside is muggy. At the bar we are plainly very old compared with the rest of the clientele. Most are young South Africans on organised bus and truck trips. They obviously don't need to sit on proper seats, as the bar has nothing but wooden planks and sand in the garden. Most have congregated around the fire; waves crash on the shoreline and a cool breeze rustles the palms. It could be paradise; it is paradise for these young things, no doubt. For us it would be paradise if only it had some nice soft reclining chairs, a flat camp area, a decent shower with running water, a place to put the rubbish... and it could be a bit cooler and less humid! Oh well, paradise is for youth and deeply tanned bodies.

We are still in a hurry; anyway we don't have time to linger in paradise. We have still not been asked for any bribes at the police checkpoints. In fact there haven't been many and they don't get up early, unlike us.

Surely as we get closer to the capital Lourenco Marques, now called Maputo, we are going to see more garages. The road is busier, the buses are more modern and the garages have diesel. We can relax a bit now, the gearbox is noisy but that's the way it's been since Yabello and even more so since Blantyre. We do worry our little heads over these noises some days. It will be a relief to get into Swaziland and South Africa. Our Portuguese is non-existent and we don't relish finding a Land Rover dealer hereabouts.

We camp outside Maputo. The place is immaculate and well run. The grass is cut, we have private barbecue covered shelters and the showers are superb. The clientele is once again totally white and South African. Most have dinghies and luxurious roof tents. All are coming and going between the South and the Mozambique coast. It's the long Easter break and the road has been busy with these foreigners all afternoon. This campsite cannot be run by anyone other than a white expat. And so it is. The gentleman behind the bar looks like he has just alighted from the 9.05 train from Victoria to Dorking and is about to enter the buildings of a well-known insurance company. But looks are deceptive here. In fact he and his wife have migrated from running a hotel in depressed Kenya to this up-and-coming stopover campsite on the newly developing coastal tourist trade of Mozambique. Things could easily get better for them, but the roads are apparently deteriorating fast and some tourists will not risk their flashy 4x4s on them.

Maputo has lost its old colonial character for the most part. The traffic is quite heavy; the atmosphere is a bit heavy too if you go down the wrong streets, and all in all we were happy to get into Swaziland. The historic railway station was immaculate, though, and obviously anyone who wants to go anywhere economically in this country must live in its large breezy capital.

The South at last

'Cyclists and pedestrians, beware of lions'

The sign at the entry to the Hlane game reserve is clear and should be taken notice of by anyone not enveloped in a motorable mesh of steel. Already Swaziland was very different from sleepy Mozambique. Just after the super-efficient border, we drove through a vast swathe of sugar plantations and then straight on into the game reserve. Later the countryside opened into a breezy hilly zone and on to the small industrial heart at Manzini.

Mlilwane game reserve, Swaziland

Reilly's old house stands on a high tree-topped hill overlooking a swathe of grassy bush with some forested areas. Afternoon tea is served in china cups from a large cosied teapot; the armchairs are worn but perfectly designed for the afternoon lethargy. In this small private game reserve one can see zebra, antelope, bird life and impatient warthogs as well as crocodiles. Mlilwane game reserve is just the perfect spot for relaxation for a road-weary traveller after the deserted broken roads of the former Portuguese colony.

Leaving the park on a Sunday morning, we passed through a sleepy Mbabane and headed on up into the high Veld country to South Africa. A shock awaited us in this remote corner of South Africa. The countryside was poorly farmed and dotted with forestry areas; the housing was very poor, much worse than Swaziland, which boasted a number of swish shopping malls and very pleasant houses, farms and country dwellings. Along the road we were shortly greeted with a welcoming warning, 'Crime Alert Do Not Stop' and later 'Hijacking Hotspot'.

Our first two hours in the country did not give a sense of having arrived in the golden lands. After Carolina, a small mixed race town, we joined the main road to Pretoria, more familiar sights with proper garages, fuel stations, supermarkets and white farmers out for Sunday lunch. Pretoria was quiet too, but at the backpackers' hostel we had to drive into the garden parking area through massive iron gates with enormous Chubb locks. Above the wall nasty razor wire indicated a not-so-relaxing environment. A notice clearly indicated the electrified wire above the razor wire and another warned intruders of the armed response likely to appear quickly should any mischievous elements be bent on attack.

Of course everyone had been saying things are not so good in the big South African cities now and it seemed Pretoria was no exception. All the houses in this otherwise pleasant suburban area were equally armed and protected. Freedom has arrived, but no one dares to enjoy it. We made a foray down to the shopping mall to find an Internet café, but that was as much as we wanted to see of town for now. In fact we decided to head straight out of town next morning to the Botswana border beyond Rustenberg and Kanye.

The man with the Afro haircut introduced himself, his appearance a certain roguish style. Born in Gaborone, he was here visiting relatives and checking out some of the bushman culture. He had just driven back from the western border of the country. His sister, not his real sister but his lifelong childhood friend, ran the small resort here in Kanye. The place was fairly deserted, apart from an Indian family and a couple of white visitors from South Africa who disappeared after their meal. The rooms here were rondavels – native circular-style – and we were pleasantly ensconced in the long grassy camping area behind.

Our new friend was in fact visiting from the UK, where he is in practice as a medical sports doctor, apparently living in Mickleover, a small suburb of Derby where Siân grew up. His unkempt appearance certainly hid his surprising career; the BMA had similarly mistaken his brash dress style at his qualification interview, he told us! A pleasant evening followed and all drinks were on the house, including the fine bottle of Cotes du Rhone that our host consumed single-handedly.

The Kalahari Desert is an elusive place. The desert seems to be lost some way to the south of the new straight and flat trans-Kalahari road. This road, a 750km modern highway, passes endlessly through low dry bush and scrub. Not long after the rains, this region was at times surprisingly green. Cattle roam across this vast wilderness. The Kalahari desert is really not a desert here, but a waterless wasteland.

Wire fences lined almost the entire route across the country. After the junction near Ghanzi, we still could find no place to camp and found ourselves at the border just as the sun set over a well-appointed campsite. And then to Windhoek…

Windhoek
The Namibian capital Windhoek is a pleasant town, set in hills with crystal-clear skies a stunning, luminescent blue. Modern buildings blend well with former German churches, civic buildings and well-tended gardens. Windhoek is a European town with street cafés, supermarkets and well-stocked shops. It's also expensive. To the north is an equally large conurbation or township; this area houses the vast majority of the Africans, it is the antithesis of the downtown area. The gap between the descendants of the white settlers, colonists and immigrants stands in stark contrast to the majority of the original inhabitants. Again we are forced to camp in a backpackers' house

surrounded by electrified wire fences and armed response notices. Going out after dark is strongly ill-advised. It's rather sad and we're glad we don't have to live here.

We set off for Etosha the next morning, but had gone barely 15km when a terrific noise erupted from the gearbox. So this was to be where it finally exploded and refused to go any further! Rather fraught, we coaxed it all the way back towards the town, passing the townships and industrial zone of northern Windhoek. Driving in low-ratio fourth gear only, somehow the Land Rover held together long enough to get into the suburbs. At a garage a white local advised us to continue into town to find the Gearbox and Diff Doctor.

We ran through a few traffic lights in our low fourth gear, barely able to stop and start, to find the doctors' clinic in the north-western part of town. There was no question of going anywhere else, so the vehicle was staying here until it was fixed or abandoned. This, of course, was now Friday afternoon, so we were taking up residence in Windhoek, like it or not. In fact nothing was to happen until Monday, once the gearbox had been removed for the second time on the trip and the third time since it had been 'reconditioned' in England.

We spent Saturday night on the coast at Swakopmund, having braved the African minibuses each way. The scenery kept us enthralled, not consciously looking at the speed the minibus was going. At least when the notice said eleven passengers only, they kept to the rule. It was Ethiopia all over again, but with more legroom. Swakopmund is a strange place, European in appearance, very quiet and somehow a bit lifeless. Some German-style buildings remain, but it didn't feel that German. They have steadily left over the years, and besides it's many years since they held sway here.

The minibus back to Windhoek was not in demand on the Sunday morning, so we were given a long tour of the black township area as they looked for customers. This, of course, gave us a rude awakening, but it did also allow us to see the township conditions from the safety of the minibus. Walking around there may not have been so easy or safe. These townships were in fact not as poor as some we had seen in South Africa. Most had electricity, small gardens and neighbourhood shops, churches and basic civic buildings. In contrast with the white areas, though, they did seem poor.

Monday morning and we were back at the 'doctors' for the diagnosis. Things were not good. Louis, the mechanic, a descendant of German settlers, gave us the prognosis. The gearbox was effectively kaput, as parts were not available from their man in Johannesburg. This gearbox was too old even for South Africa. It seemed some colleagues might have a slightly less old gearbox that should fit, but we wouldn't know that until the next morning.

Gearbox and Diff Doctors
The workshop was run by three Germans, and a more efficient garage you would never find. These fellows had an amazing pride in their workmanship and engineering. When the gearbox was installed it was placed a few inches further back. This needed new mounting points and a change in the length of the propshafts. That job was done elsewhere, again by white engineers. The shafts returned next day repainted in glossy black – a work of art. The general labouring was done by some skilled and some unskilled blacks. This of course immediately presented us with the sharp contrasts of the country, literally in black and white. The whites drove immaculate 4x4s, while the blacks caught the minibus home. Each group in their own way was welcoming and helpful in the extreme, but what frustrations and tensions lay beneath the surface?

The whole pattern of social structure, inter-communal tensions and the difficult mixing of two cultures was as pronounced here in Windhoek as in Pretoria, but on a smaller scale. We could not help thinking that it could all go the same way as Zimbabwe.

We retired to the nearest backpackers' hostel, the Chameleon, cooked our dinner and sank into lethargy in front of a large TV showing videos. Tonight they were showing Seven Years in Tibet. We hope we are not about to spend seven years in Windhoek. Next morning the plan had changed and so had the estimate for the cost of the repair. It almost seemed to make more sense to abandon the vehicle and the trip and take a bus to Cape Town just 1600km south. What to do?

The cost was far more than the value of the vehicle. Of course we had had this vehicle for fourteen years, it had been to Africa four times and the Alps as well. It seemed completely mad, but we didn't want to 'give up' just yet. The show had to go on, even if it was for only three more weeks. Except that we still didn't have any news from the shipping agent about the date of the voyage to Southampton.

Ten days after arriving in Windhoek, we were mobile again, though our bank account was somewhat depleted.

Dead Vlei near Sossusvlei, Namibia

The dunes were a stunning red colour, quite the most red we had ever seen. Dead Vlei was somewhere ahead across a low dune, hidden. Odd-shaped, virtually lifeless trees dotted the landscape. The trees stood like ghostly reminders of gnarled goblins in a sea of blisteringly white salt pans, totally surrounded by imposing red dunes. Namibia was proving much more exciting for its wild landscapes, its colours, its rugged mountains and its tranquil vistas. One could only admire the original settlers who opened up the terrain. The camping was great and the starry skies magnificent. Finally we could sleep on the roof of the Land Rover. The southern stars were so amazing we actually lay awake for hours watching them drifting across the night sky. It was great to be on the road again, and hang the expense.

At Helmeringhausen, a fine African name, we stopped at the famous hotel for mid-morning apple strudel. The best we'd had since the one two days before at Solitaire! We diverted west to Luderitz, said by guidebooks to be a jewel of a Bavarian village on the desert coast. Well, not quite, but Luderitz certainly is a strange place with its windy wild seas, its German houses and its apparent dying history.

Fish River Canyon was stunning. The campsite was modern and pleasant, occupied by a number of South African overland trucks, most with not enough clients to make any money. It was 28 years ago that Bob had first visited the canyon.

Sadly we had to leave Namibia the next day and, after our extended stay in Windhoek, we were forced to miss Etosha, the Angolan border area of the Ovambo tribes and the Skeleton Coast. It always happens, there is always another place to be revisited...

German architecture in Luderitz, Namibia

A hazy sunshine greeted us as we parked by the seaside. The car park had an unusual red brick surface and a couple of tour buses pulled in. This was the place to come for the view. At this point we were more troubled by our lack of communication with the shipping agents about getting the Land Rover home. The internet had crashed at the café and we seemed to be in big trouble. It's a pity that we had to take in the enormity of the view from the car park so hurriedly. Table Mountain was clearly in the picture, if a little hazy. This was the Cape, the symbolic end of the trip.

We had made it, but nobody else knew it. American tourists poured from the coach, another day; if it was Wednesday, it must be Cape Town.

A hazy day in Cape Town

South African woes

We passed depressing townships full of vibrant people, slums full of cardboard boxes that people call home. What will become of the country? We visited the luxurious estates and a family friend who lives in Constantia, where the elite live behind electric fences. Even the new black elite are in prison of a sort. The black population naturally expects more and more, but their numbers are expanding too fast, AIDS is a massive scourge, and crime seems everywhere. Many of the whites are very rich; some have worked hard for it too. But some whites are now poor and destitute, living in campsites that we shared with them. These whites are as destitute as the underpaid blacks. It's a tangled web. The genie is out. Africa has shown us every human facet, diversity and too much to ponder. We of course are privileged to be able to sail away from all this.

The trip up to Durban across the Karoo was exciting. We visited the eccentric little settlement of Matjiesfontein, a time warp where brave colonialists transported a piece of old England to the vast mountain plateau of southern Africa. The Land Rover played up again, with starting problems and a leaking clutch master cylinder. We were as always slaves to our vehicle. Ultimately, we depended on it for our lives. It had got us across a continent; it was magnificent. It epitomised our apparent wealth, it was the centre of some envy, the cause of much frustration, it was our home and yet again it lived another day.

10 May: Durban
'A ship is leaving for Zeebrugge on 12 May, it has space and yes you can put your vehicle on it, why not indeed?' explains Alex, an Indian, the local shipping agent. Yes, this ship was going to Southampton.

****

After four months we finally felt a sense of relief, mingled with a sudden sadness, as we rolled into Chichester once more.

Chichester Cathedral, West Sussex, England

Travellers' Tales

Following our trip to Africa, we were asked to write the 4th edition of the Bradt Africa Overland guide. It was quite a task but relished with enthusiasm. All those stories and experiences could now be of some use to fellow travellers. Perhaps that was our innate purpose in life after all? Maybe not.

Island Paradises: 2004–5

Just after submitting the manuscript for the Africa Overland book for Bradt, they asked us if we had time to do some updating work. There would be no real payments but enough money for a trip or two. Why no one else had jumped at this opportunity already was a little surprising. There didn't seem to be any reasons for anyone to fear the destinations.

Maldives: December 2004

The first option was to go to the Maldives and update the manuscript and maps for a new edition. Of course we were never expected to visit every island – there are only a hundred or so with some tourist infrastructure to check out. Our principal task centred on the main island, Male. It was big enough to take up a few days, with lots of hotels, restaurants and sights to see. In the event we were taken to a number of the closer islands to get a better insight into the needs of visitors. The island of Giravaru was one that we were able to stay at; its underwater sea life was a revelation to us land lovers. Just two metres from the pier a myriad of multicoloured tropical fish lived, where the reef dropped into the dark abyss. It was a really fortunate opportunity that we had never expected, completely outside our normal zone.

A couple of weeks later, we were flying back to the UK from our work in Nepal and heard of 'a big wave' on the news. This was before anyone had ever heard the word 'tsunami'. Knowing how low the islands were, with no high places to run to, we were horrified to think of the terror the inhabitants and holiday-makers must have felt at the sight of this destructive freak of nature.

Local dhoni in the Maldives

South America: February–April 2005

The plan was a little ambitious for the time restraints. We flew to Buenos Aires and took a ferry across the River Plate to Colonia in Uruguay before heading south.

An old car in Colonia, Uruguay

Flying as far south as it's possible to get without going to the Antarctic, our journey really began in Ushuaia. The weather played ball and the scenic charms of Ushuaia were a surprise. The seas were balmy and calm, reflecting the rugged grandeur of the backcountry and the magical pure blue skies. Across the border in Chile is Puerto Valdez, another Scandinavian-looking place with climate to match.

Perito Moreno glacier, Argentina

Our luck with the weather failed at the Torres del Paine National Park, so we didn't get to trek there. Not far away is the amazing Perito Moreno glacier, an ice river of immense proportions. At the time of our visit the ice was exceedingly close to the viewing areas. Great shards of green-tinged ice periodically tumbled down into the icy grey waters of the lake. Apparently the snout of the glacier is sometimes a considerable distance from the shoreline, needing a boat to get as close-up as is safe.

Further north from Calafate we were again blessed with good skies. The trekking around the base of Fitzroy and Cerro Torre was as inspiring as any in the Alps. Luckily the path was quite easy, as we had not brought our heavy hiking boots. We did the circuit below the two peaks twice, as the weather was even better the second day, but our shoes suffered ignominiously.

Fitzroy Peak, Argentina

Siân playing the organ in Trelew, Argentina

Travelling north, our plan took us to Trelew and the Chubut, the Welsh area of Patagonia. Being from Wales, Siân wanted to see the land where some of her compatriots had migrated to so long ago. It's a long way to get away from the English!

From Puerto Montt we flew all the way up to Arica. The bus climb across the Andes to La Paz proved as exciting as the route taken by Bob on his earlier journeys. Surprisingly the effect of altitude wasn't as bad as one would expect in Nepal.

The peak of Illimani watches over La Paz, Bolivia

Bowler-hatted ladies in a street market, Bolivia

La Paz had outgrown its great bowl-like canyon and spread up onto the high altiplano above. Modern-looking El Alto is like a new satellite town to La Paz, but the old city has retained its charm. Colourful people still throng the square of the San Francisco cathedral, the colonial quarter is as vibrant as ever and the steep, narrow streets just as knackering to climb. We stayed at the atmospheric Hotel Torino, a favourite of the budget travel brigade until recently.

We diverted north to explore Lake Titicaca and then sped on south by train to Uyuni, where a vast expanse of blindingly beautiful salt lakes shimmer in the sunlight. Further south is Tupiza; a great side trip here is to head out to the wild desert of the southwest of Bolivia. Startling wind-cut canyons, placid reflective green lakes, breezy purple lakes peppered by flocks of flaming red flamingos, smoking, ominous-looking smouldering volcanoes and multi-coloured hillsides combine in one mass of adventure and natural beauty.

Along the way we found occasional internet cafés, sufficient to download and work on emails from Bradt regarding the editing of our Africa Overland book. Once in Argentina, our route took us to Salta, to sample the extraordinary dulce de leche with a very welcoming family friend Sheila Misdorp. From there it was east on a long overnight bus trip to Posadas.

Iguasu Falls, Argentina/Brazil

After a quick visit to the Jesuit ruins near Encarnacion in Paraguay, we continued to Iguasu Falls. Cheating now, we flew to Rio and then back to London.

Panoramic views over Rio De Janeiro, Brazil

Cape Verde: May 2005

A few weeks later, again in search of new data and updating for another Bradt guidebook, we found ourselves on a plane – several planes in fact – to somewhere else we had never expected to go. The collection of islands that make up the country are as diverse as the imagination will allow. One is a desert, another a volcano, another a dry, undulating scrubland. Another is as mountainous as an island will permit while another is a quiet backwater that few have ever heard of. The Cape Verde Islands are now quite a popular destination, but back in 2005 they were mainly the preserve of European package tourists. All our air tickets were taken care of by the national airline, but after that we of course slummed it with lodgings and food. But this was another once-in-a-lifetime chance to get to another place that was almost off the map.

Sao Nicolau, Cape Verde

Mount Kailash and Tsaparang: October 2005

The trip to Mount Kailash is never an easy option. The planning is Herculean, with bureaucratic obstacles as high as the Tibetan plateau. Getting to the isolated peak is problematic, with high snow-prone passes and poor roads. When one arrives at the base of the peak, there is a very high pass to cross to complete the circuit. Of course this stuff is what adventure is all about!

> *The power of such a mountain is so great and yet so subtle that, without compulsion, people are drawn to it from near and far, as if by the force of some invisible magnet; and they will undergo untold hardships and privations in their inexplicable urge to approach and to worship the centre of this sacred power.*
> **The Way of the White Clouds,** Lama Anagarika Govinda

On the drive in, the weather proved the main hazard, as a snowstorm provided too much excitement for a couple of days. The town of Saga is showing the last vestiges of its Tibetan heritage, as its new incarnation sprouts all around in the form of new Chinese hotels and streets that suit the ambience of a modern world. The only way to keep warm in our ageing hotel was by standing under the infrared lights; the sort of lights that were used by chicken farmers (like Bob's father) in the UK to keep the chicks warm. A day was lost in this cold storage lodging, but at least it helped us to acclimatise. The road journey took us via the desolate, often sand dune-lined ranges of the Trans-Himalaya via Paryang to Lake Manasarovar. The holy lake revered by Hindu and Buddhist pilgrims has many moods: some light and airy, others dark and brooding. Our small guest house had a fabulous view of Mount Kailash from the outside loos. We enjoyed this view quite a lot!

Before the trek around Kailash, our plan was to visit the remains of the ancient kingdom of Guge to the southwest of the peak. Taking a remote dirt track, our route went close to the lost remains of the Bon citadel of Khyunglung, although our crew seemed oblivious to the significance of the place, so we weren't sure if we actually saw any of the historic site. No matter, the scenery was unworldly all the way through Dawa citadel and the Mangang valley. The whole region is eroded into contorted troglodyte turrets, chasms and defiles that defy description.

Tsaparang Citadel, Guge, Tibet

Our main destination was Toling, where the 11th-century revival of Buddhism took place under the Indian sage, Atisha. Long after Buddhism in India had lost favour to Hinduism, Buddhism began a long and slow renaissance in its often strange but colourful Tibetan incarnation. This Buddhism was to cross back over the Himalaya to Nepal, Bhutan and Ladakh in the form that is observed today in those areas.

In Toling we visited Yeshe O Monastery, the main remnants of the great revival. Not far away is Tsaparang, perhaps one of the most mysterious and exotic places on earth. Its remote locality, the startling citadel and sensational rocky outcrops really give meaning to the much-touted concept of Shangri-La. This was no Shangri-La of comfort, but the edifice is nothing short of mind-blowing, a dreamlike apparition.

Deep within the ruins is found some of the most astonishing Buddhist art in the world, although sadly the Red Chinese desecrated some of it during the Cultural Revolution. Fortunately enough remains here and in Toling to entrance any visitor interested in mystical antiquities.

Indian Himalaya including Kamet above Toling, Tibet

Climbing over a high pass (5100m) back to the main road to Mount Kailash, a vast panorama unfolded. The Indian Himalaya, including the stunning spire of Kamet peak, flowed along the southern horizon. Lost below was a vast plateau, cut, riven and dissected by thousands of eroded turrets, towers and mysterious canyons.

The trek around Kailash was fascinating; being late in the season there were hardly any other visitors, save for a group of wild-looking Tibetans. The trail is relatively flat for most of the way as the route carves its way along the canyon. Circling around to the north, the trail climbs steadily to Driraphuk Monastery. The accommodation was very basic – the floor. Food was no better – noodles. Sleep? Well, that was pretty hard to come by too. This is a pilgrimage, it is supposed to be uncomfortable. Towering above is the gigantic and almost sheer north face of Kailash; an overpowering scene to inject any forlorn trekker with divine inspiration.

Literally and figuratively, the high point of the route is the prayer flag-dominated Drolma La Pass (5630m). For most people, getting to the pass is a big challenge, but the rewards are great. Fortunately the best views of the circuit are just before the final slopes, where the northeast walls of Kailash are a stunning vision. After descending past the bleak Gauri Kund lake, the way is easier and the loop back to Darchen very pleasant.

Siân & Bob at Mount Kailash, Tibet

It is said that those who complete the circuit of Kailash are divested of all their past sins. To be completely cleansed and absolved, one must complete 108 circuits of the mountain. That's us done for, then.

And we are still to find any definitive answers to the mysteries of life.

More guidebooks

Following this trek, through contacts with our old friend Kev Reynolds, a prolific Cicerone author, we were asked to do a guidebook to trekking around Mount Kailash by the British publisher Cicerone. It seemed that this new path was more than a passing whim.

Ladakh and Spiti: February 2006

Shortly after this trip we were invited by Rama Tiwari and Joanne Stephenson to produce a guide to the monasteries and culture of Ladakh. The research was done in February, hardly the optimum period to visit the high snowbound region. Anyone who has been to Ladakh will know how captivating it is, at any time of year. The monasteries are impressive, many clinging to sheer cliffs. The people are friendly and the Buddhist culture is inviting. The landscape is stunning, with high desert plateaus and endless mountain ranges cut by luminescent ribbons of turquoise water in deep river valleys. Ladakh is synonymous with adventurous travel.

Historically linked and relatively close to Ladakh and the Guge region of Tibet is Spiti. Getting into the mountain fortresses is incredibly scary. The road from Shimla cuts across high passes and then dives and ducks along the Sutlej valley, the same river that drains from near Tsaparang. Often the road is cut into the sheer-sided canyon thousands of feet above the river. The bus drivers 'trust in God' but we mere mortals with some scepticism may not be so confident.

Higher up the landscape is again Tibetan and dotted by more historic remnants of the greater Guge Kingdom. The Tabo monastery houses perhaps the world's greatest Buddhist imagery and artistry from the 11th century and later. The Dalai Lama has visited on a number of occasions and probably feels as close to his ancestral home here as is possible. Near the administrative town of

Kaza is the dramatically located Ki Monastery. The old citadel of Dhankar is as impressive as that at Tsaparang. The whole Guge Kingdom is so little known that its mystique is sure to remain forever, barely accessible to this day.

Chemrey Monastery, Ladakh, India

Bhimakali Hindu temple, Sarahan en route to Spiti

Canyons and Deserts: April–May 2006

In the spring Cicerone asked us if we would like to go to the USA to update a guidebook to the Grand Canyon. It was a chance too good to reject, even though there would be insufficient funds to cover all the costs. The trip was a revelation, wiping away all the negative images that one gets about America and Americans.

We flew to Calgary to visit Bob's sister Chris and husband Phil and then hired a car to drive south. It was winter in Montana and Wyoming, but sunny enough to appreciate the Tetons en route. Bryce Canyon and Zion were the first parks to explore. Chris and Phil flew down over the Easter weekend to join us. Snow still dusted the canyons but the walks around the two parks were breathtaking.

Chesler Park, Canyonlands, Utah, USA

We were all blown away by the scenery in both parks. Later we drove on to Capitol Reef, Arches and Canyonlands national parks, where we did some amazing day hikes, making copious notes for a possible guidebook. The well-designed campsites were a dream for those who love the great outdoors.

The Grand Canyon was the main focus after this and it surely can never disappoint. It was easy to see why some trekkers get into difficulties climbing back up from the depths of the canyon. Before heading back north to Canada, our route took us to Death Valley, Sequoia, Yosemite, Craters of the Moon, Yellowstone and Glacier National Park. What a fabulous trip this turned out to be.

Korean surprises: December 2006

The 2006–7 trip into North Korea between our seasons in Nepal was organised in Kathmandu, by email through a travel agent in Seoul. At the time a 3-day tour to a special tourist enclave was permitted by the North. The short entry point across the DMZ was north of Soraksan. The group was based in a couple of hotels in a fenced area near the village of Onjong.

The famous Kumgangsan, the Diamond Mountain region in North Korea, was the main objective of the trip. The attractions of Guryeong and Manmulsang provided some surprisingly rugged trekking. A stranger situation could hardly be envisaged, with North Korean guides and guards helping mostly South Koreans to visit this 'holy' mountain. From the top floor of one hotel it was easily possible to look into the real North Korea. The local village had no electricity, houses were medieval-looking and there were no cars or trucks in sight.

Onjong, North Korea

Guryeong trek & waterfall, North Korea

Beopchusa Buddha, South Korea

Changing the Guard, Seoul, South Korea

Travelling around the South was easy; the buses ran on time, were comfortable and good value. The number of picture-postcard Buddhist shrines and temples found across the country amazed us. The highlights were climbing Soraksan Mountain in the north, the shrines of Beopchusa, Bulguksa, Golguksa and Maisan Tapsa, as well as the traditional Andong Hahoe village. Three towns seemed to share similar-sounding names – Kyongju, Gyeongju and Chongju – but each held its own surprises.

We didn't really take to the food (especially the kimchi) though! The under-floor heating in the hostels was a great bonus for this midwinter trip. We celebrated Xmas Day in Seoul with a tasty Korean burger.

Before returning to Kathmandu, we deviated a little to visit the Malaysian part of Borneo and the fascinating kingdom of Brunei. It was interesting to find that Bob was a senior citizen at only 55 and had reduced-price entry to the national parks!

Karakoram Highway: October 2007

For a change we took a different route to Nepal for the autumn season, this time utilising a new flight to Dushanbe and back from Kathmandu via Delhi for a trip along the Karakoram Highway. Unfortunately the Tajikistan visa issued at the airport was limited to 5 days, due to a high level conference of the ex-Soviet states headed by Vladimir Putin.

Local sellers by the mosque in Dushanbe, Tajikistan

Having hurriedly visited the capital and Kojand in Tajikistan, we then spent a whole day sitting on the ground in a cotton field on the border waiting to get into Uzbekistan because of security issues concerning the same conference. When they finally let us into the country, it was after dark and customs was closed. The historic city of Kokand proved to be an interesting stop with plenty of blue tiled mosques and ugly Soviet-era concrete.

With so many strange and inconvenient border delineations after the independence of the Central Asian states from the Soviet Union, memories of the conflict in the Ferghana Valley between Uzbeks and Kyrgyz were still raw.

We finally crossed into Kyrgyzstan after a big hassle from the Uzbek customs. They did not want to let us leave without producing a currency document that we did not have, because customs had been closed when we entered the country. The second city of Kyrgyzstan, Osh is another strange mix of 'exotic Islam' and 'ghastly Soviet', but fascinating for that. The drive to the town of Saray Tash proved very dramatic, crossing a high range of the Tien Shan. Saray Tash was little more than a few yurts and some basic houses, but the views to the south were sensational. From here a rough bus ride led to the Chinese border and a shared taxi to Kashgar.

The Khan's Palace, Kokand, Uzbekistan

Id Kah mosque, Kashgar, Xinjiang, China

Kashgar had changed dramatically since our first visit in 1985. The old Id Kah mosque remained but it was now surrounded by a swish new Chinese plaza. Wide streets had been blasted through the old mud-walled areas to the south. A token area of the old city remained, but the atmosphere of intrigue and history had gone. The ghosts of the Great Game, between the vying empires of Russia, China and Britain, were well laid to rest.

The climb into the Karakoram was spellbinding, passing the great peaks of Kongur and Muztagh Ata set like jewels above Karakol Lake. Crossing into Pakistan was easy and the road dramatic. Switchbacks and hairpins were taken at Formula One speeds in first a jeep and then a rickety taxi down to the Hunza Valley.

Kashgar 2007

Karakol Lake & Kongur peaks, Karakoram, China

Eric Shipton, the climber, explorer and former consular officer in Kashgar, once said that the Hunza Valley is one of the most magnificent mountain vistas in the world.

And few will argue with that.

Karakoram Highway views, Passau, Pakistan

Hunza Valley, Pakistan

With trouble on the road below Gilgit, we decided to fly down to Rawalpindi. When Benazir Bhutto was shot the same day, we decided to take the next flight directly from the airport to Lahore and a taxi to the Indian border and Amritsar. It was a sad end to the trip and the last time we visited Pakistan. Ironically Siân had been at Oxford at the same time as Benazir Bhutto, although their paths rarely crossed.

Benazir Bhutto was assassinated not long afterwards.

The Overland Bug Bites Again

You would think our long trip in 2004 would have quenched the desire for any more overland. Well, it did take another three years to fester in the system.

West Africa Overland: January–April 2008

This trip was unique for many reasons. One was surprisingly that it didn't cost us a penny. By some amazing fluke, we were asked to do some research in Morocco, Mauritania and Mali for the well-known trekking company, then called Himalayan Kingdoms. We had met Steve Berry, the director, at one of the London travel shows. As well as planning his own overland trip across Asia, he wanted to find some new destinations in Africa.

The Kasbah of Ait Benhaddou, Morocco

A shocking road in Guinea

The trip took us down the long Atlantic coast of Morocco and Western Sahara to Mauritania. The road was sealed all the way now to Nouakchott. Heading into Senegal, the border was a bit of a hassle. Touba is one of Senegal's most religious sights, a large blue-domed mosque requires western visitors to act with decorum. The road into Guinea was a shocker, a killer of shock absorbers in fact.

Siân and Madame Raby, Labé, Guinea

At Labé we stayed with Madame Raby, a welcoming and charming 'mama' on the budget travellers' circuit. Actually there were only a few travellers in Guinea; most were learning about its fabulous musical heritage or trekking around the Fouta Djalon highland villages. Of course Guinea has had its ups and down politically, but after some terrible years it's a lot more placid now. Guinea will grow on you the longer you stay. Don't expect much comfort, but its people are as friendly as any across the continent. Isn't that always the way with the common folk of any country in the world? Why do people have to suffer tyranny, dictatorship and wars promulgated by those who are addicted to power?

Heading into Sierra Leone, would we find any good roads, hotels or blood diamonds?

Freetown cannot be described as an inviting paradise by most, but delve a little deeper and there is always something memorable to record. Along the coast are some of the best beaches in West Africa and only a handful of visitors. It's sure to be 'found' by developers soon enough. The war-torn country still bears many scars from the war, but it's on the up in many ways. Yet there are signs of the conflict in surprising quarters!

From the diary
The two men sat drinking beside the pool, as dusk settled over the country hotel just outside Bo in Sierra Leone. They looked like missionaries; who else would be driving around here apart from us?

'Sit down; have a beer.' It was abruptly clear that the two guests were not missionaries. One had been living in the country for most of the last dozen years. The other had arrived from, among other places, Guinea. 'With the war over, where is all the promised development?' we asked. 'Well, the war is technically over, but come into town (Bo, which was the rebel RUF headquarters during the struggles) and I'll show you where to buy an AK47 for a couple of hundred dollars.'

In the streets of Bo there are many shops advertising 'blood' diamonds for sale. So how was the road to Liberia, we asked. 'It's not good, but it's fine during the daytime; the villagers will help you and look after you if you have a problem. Just don't break down at night! ... And don't stop to help anyone along the road – it might be a set-up,' they said. Very encouraging!

We had already done that in Guinea, when a taxi-brousse full of desperate-looking locals had broken down and needed engine oil. For us, not to stop would have seemed harsh. It's a gut reaction thing; react as you see fit for a given situation and be wary.
'So what is your mission in the country?' they asked.

'Just tourists,' we replied. We started to feel rather silly. 'Tourists in Sierra Leone!' one exclaimed. 'I thought that had stopped years ago! Even then it was only along the beachfronts of Aberdeen and Lumley, which the Chinese have bought up recently.'

Our hosts were clearly wise to many things that we could only imagine from media hype of the past and the film, Blood Diamond, which portrayed the country in a poor light. So far we'd had a pleasant time in the country – the coastline near Freetown is spectacular, a paradise lost. Even frenetic, sweaty, overcrowded, dilapidated Freetown has a certain disintegrating charm.

Freetown, Sierra Leone

The drive on was exciting, but only because the route was another fun, off-road, dry muddy lane to the Liberian border. The village people were very friendly and we didn't break down. The first people to greet us at the Liberian border were Pakistani peacekeepers.

The UN in Liberia
Being greeted by a large white tank and soldiers with blue helmets was both welcoming and intimidating when we arrived at the Liberian border post of Bo Waterside. 'Salaam Aleikum.' The Pakistani soldiers seemed eager to talk. 'What is your country, madam memsahib? Pakistani cricket team, it is good, yes sahib?' This was the Pakistani Battalion, nicknamed Pak Batt, motto 'Brave and Brisk'.

So what is Liberia like?

The capital Monrovia shows only a few signs of the wanton destruction of the war. The markets are stuffed with imported goods; huge generators hum all day. The restaurants are expensive, catering to the armies of UN advisers, and the hotel prices match their salaries. Business people are making the most of the new investment opportunities. However, the rainforest in the countryside that we saw is mostly devastated. The main roads are fair tarmac, but electricity is not widespread. The rubber trees are still dripping and the Firestone Tyre Company is working again.

Leaving Liberia upcountry for the backroads of southern Guinea, we were in the land of the UN Ban Batt, the Bangladeshi Battalion. Liberia is a great place. It has the first woman president of any African nation; perhaps that is reason enough for hope!

The main cart track 'road' from Sierra Leone had been a shock to the system, and escaping unscathed from the diamond smugglers of Bo was a relief, but getting this far was very exciting. We really hadn't thought it would be this easy to visit the country, nor had we guessed how safe it would feel. How many AK47-wielding rebels would we see? You really can't judge from any government travel website just how sensible it is to make these journeys. Yet on the whole trip we never once felt intimidated by anyone, even on this diversion to the recently war-torn former American colony. Optimistic, happy-looking people greeted us everywhere. We never saw any gun-toting soldiers of any persuasion, except the UN.

Bombed-out central Monrovia 2008

The next part of the trip took us back into Guinea and then Mali. The pearl of West Africa, according to many travellers, is the mud-walled town of Djenné with its superb Sahelian architecture.

Although no stranger to much of Mali, Siân had never been to Timbuktu. With a more reliable engine in the Land Rover this time, it seemed worth the risk of crossing the desert from Douentza. In the event the road wasn't as bad as feared. We made it to the Niger River in one hop, and shared the narrow ferry with four large camels on each side of the Land Rover.

Djenné Mosque, Mali

The Hand of Fatima outcrop, near Douentza, Mali

Timbuktu is a city of explorers, mysterious blue men of the desert, fairytale mud mosques and stories of former untold wealth. It hosts the tangible melancholy of a former glory. Timbuktu today is an enigma; the incessant wind blows plastic bags across a sand-laden street, grubby children follow a stranger and modernity still struggles to impose its antiseptic character. Timbuktu does not disappoint.

Above: En route / below: Djinguerey Mosque, Timbuktu, Mali

Alexander Gordon Laing, Heinrich Barth and Rene Caillie all reached the city of Islamic learning, mostly in disguise, at great personal cost and a great deal of physical discomfort. Laing was the first, but he was killed on his journey home. Caillie made it back; Barth stayed for years and survived his journey. The houses of these great adventurers, where they hid or lived, still exist within the sandy old city. Why would anyone, particularly an 'infidel', want to come to Timbuktu?

Having first reached the city in 1978, I was apprehensive about the impending return. Familiar sights remained, though – the impressive Djinguerey Mosque now given a new paved access path, the Sankore Mosque extended in tasteful style and the house of Laing still signposted but with at least 1m of sand removed from the street.

Bob in Timbuktu, 1978 and 2008

Mali

In 2004 we stated in the Bradt guide:

'Where the early explorers awaited uncertain fates on their return journeys, those reaching the fabled city of the Mali Empire now have a good chance of getting back home afterwards. The internet of course has arrived in Timbuktu, but overland access is still rough, wild, lonely, rugged and, yes, achieved with a significant degree of physical discomfort. Long may it last!'

It was too soon to celebrate. Now in 2015 we all lament the chaos of northern Mali, where the Tuaregs still strive for a better deal from their southern rulers and IS is penetrating further into the Sahel. Oh what a wonderful world!

Huge Baobab trees litter the Sahel

From Mali we continued south through Burkina Faso. Burkina Faso seemed a very peaceful place, although some of the past presidents were pretty dreadful. With some stability the country seems to have blossomed into a very exciting tourist destination for those visitors who seek something different. In Togo, we were driving south past endless roadside sellers and producers of sacks of charcoal, in a dense fog of carbon smoke. Who is talking about the carbon footprint here?

We had hoped to continue through Benin and Nigeria to Cameroon, and eventually on to Angola and South Africa. However, Angolan visas were almost impossible to get, so we reluctantly started the long drive back. This turned out to be a blessing in disguise when a tendon in my right hand swelled up and finally snapped at the border between Mali and Mauritania, near the point where several French tourists had been murdered two months earlier.

Then a horrendous oil leak developed, a black dotted line marking the route behind us. The chief mechanic at the main dealer in Nouakchott recommended that we just buy as much engine oil as we could carry and continue on to Morocco before trying to fix it! A nerve-wracking journey across Mauritania was followed by a backstreet engine repair in Ouarzazate.

And finally, the freezing cold of northern Europe and urgent surgery to weave together the tendons in my right hand.

The Grand Mosque of Bobo Dioulasso
Burkina Faso

Tamberma conical houses, Togo

Caribbean cruising: December 2008

The idea of a cruise around the Caribbean was not high on our list of to-do trips until the company Ocean Village came up with some very cheap-and-cheerful cruise packages at the end of 2008.

Dutch-influenced Curacao in the Caribbean

A three-week trip that visited almost all the islands of the Caribbean was on special offer. Again it was a chance not to be missed, because flying between the various islands by plane and staying in hotels seemed horribly expensive and way beyond our budget. On most of the islands it was quite easy to take local buses to explore the sights, so it was a bit like any other trip but without the hassle of finding a room and food. No bread and bananas on this trip!

Eastern Europe Overland: January–March 2009

In 2009 our planned winter overland trip along the old classic Asia route was aborted, after an attack in southwest Iran and some pretty dire stories coming from terrorised Pakistan put us off. We did get to explore parts of Eastern Europe as compensation, but in the middle of winter it proved a chilly adventure, with jerrycans of 20 litres of water freezing solid in the back of the Land Rover, and bananas turning black. Romania, Bulgaria and the curious backwater of Moldova were enough to excite us, though.

Historic Brasov, Romania / Chisinau, Moldova

313

Indian Ocean island delights: May 2009

We were left high and dry in late March with nothing to do until the end of June. We had just had a couple of good seasons in the Alps and decided to blow some of the money on a short trip to the Indian Ocean. We found some good deals on the flights, but it was a seriously expensive destination despite using the cheapest guesthouses, a few dodgy places and eating more bread and bananas.

Anse Takamaka rocks beach, Seychelles

Flying via Nairobi, we headed to the Seychelles. It was as glorious as the brochures make it out to be, but we were staying in a family-run lodge for four nights. It was cosy but not luxurious. We hitched and took whatever local transport existed to see as much as possible.

Port Louis old and new, Mauritius

Mauritius seemed more like a real place and a lot of the country was devoid of tourist hordes. We stayed at a cheap Indian guesthouse in the south of the island, taking daytrips across the island. It was as basic as any found in India, but quite adequate for the few days we stayed. The capital Port Louis was a delight to explore and the west coast had some great beaches backed by quite impressive peaks.

Cilaos valley, Reunion

Our favourite island was probably Reunion, the French territory in the Indian Ocean. With three major valleys cutting into the central volcanic heartland, there was no shortage of trekking and superb scenery. The gites weren't so grossly expensive and here and there were hostels. On the east side the Salzie/Hellbourg valley was deep, green and picturesque. The southern area was characterised by lava flows. On the western side is the Cilaos valley; another deeply incised canyon area with panoramic vistas. The highlight was the ascent of the Piton de la Fournaise volcano across a great crater of lava, twisted and contorted into coils of metallic swirls. The summit crater rim was very unstable-looking; signs warned of the risks but the gate was open…

With Madagascar suffering a civil war at the time, we were unable to visit and had to fly over to the almost-forgotten outpost of Mayotte. This is another French possession, although they probably aren't as keen to hang on to it as they are with Reunion. It's a pretty place, but there's not much happening. The locals seem to be in fear of being abandoned by the French in the foreseeable future. There was a national bus strike, so we couldn't get out of the main town Mamoudzou.

The final island on the itinerary was Comoros. The population here is Muslim; there are plenty of mosques to see and some very atmospheric alleys in the capital, Moroni. It's another intriguing island with little public transport and not much food either. The colourful women here paint their faces with white or cream paste, but few wear the chador. Some of the older buildings were constructed of coral, making it look a lot like a rundown version of Zanzibar. Not that we have ever been to Zanzibar to confirm this. The centre of the island is volcanic and remote. Back in Nairobi we did the usual rounds of old friends and stayed at Upper Hill, where Jesse was still in charge.

On this trip Bob had to be careful not to eat greasy food, since he was due for a gall bladder removal operation on our return to the UK. Luckily the bread on the trip was tasty. It hadn't been long since a previous operation to have some bone removed from his nasal passages to alleviate chronic sinus problems. What's left to remove, a brain?!

Above: Moroni harbour, Comoros / below: painted lady, Comoros

Northern South America backwaters: January 2010

Tucked away on the northeast coast of South America, Guyana, Surinam and French Guiana are hot, sweaty destinations, rarely on the traveller's radar. Our two-week adventure here at the very start of 2010 gave us a fascinating glimpse into some little-known places.

Georgetown statehouse, Guyana

Very Caribbean in character, Guyana is an easy place to get about, along the coast at least. Adjacent Surinam is another place whose interior is hard to access; the Dutch colonial influences make it particularly interesting. The capital Paramaribo has some gems of colonial architecture and it's an easy-going place for visitors.

French Guiana is well known for the space-launching facilities at Koroue and the offshore prison made famous by the novel Papillon. It is perhaps less well-known for being a former penal colony.

In fact the capital Cayenne has some picturesque quarters, and a couple of budget hotels make it affordable. Very sleepy and not one of the better French civil service postings, Cayenne is as hot as the pepper of the same name.

As for the interior, it's wild, untamed and pretty much off-limits, with little infrastructure. All the taxi and bus drivers seem to be competing in a Formula One Grand prix race while talking on their mobile phones and to their passengers at the same time.

Elegant mansions in Paramaribo, Surinam

Cayenne, French Guiana

Africa Overland: 2010

By 2010 a new edition of our Bradt Africa Overland was imminent. There were still some countries on the continent we had not visited, so it was another opportunity to seek out new routes and places.

This time our plan was to get to Egypt via Turkey, Syria and Jordan. This of course took us back to so many favourite haunts in the Middle East. Winter was pretty cool in Turkey in January, especially with the rattling doors of a Land Rover that was designed to never fit properly. With a little more cash this time, we stayed in a cosy cave room in Goreme. It snowed all the way across the Tarsus to the coast.

Hagia Sophia, Istanbul, Turkey

At the magnificent crusaders' castle of Krak des Chevalliers, it snowed overnight and the hostel owner had locked all the doors. There was no way out except through the first floor window. Fortunately the war in Syria had not yet started, so we did not feel threatened, just inconvenienced.

We had hoped to get into Lebanon, but there seemed to be all sorts of rules about left-hand-drive cars, diesel vehicles and extra high taxes to re-enter Syria. In the event we just trolled up to the border later near Damascus.

We visited the Christian churches at Seynada and Maalula. Later we would see these places splashed all over CNN, BBC World and Al Jazeera, as they fell under the terror of the war. In Damascus the slightly claustrophobic hotel had good parking and colourful owners, so we stayed an extra day exploring the great bazaar, the Omayyad Mosque and eating the great local hummus with mint tea.

The locals, as always, were extremely welcoming.

Krak des Chevalliers crusader castle, Syria 2010

Maalula before the war, Syria 2010

We left for Jordan, little knowing that a few months later the whole of Syria would be engulfed in the deadly war that seems to show no sign of abating. We wonder what happened to all those friendly people and the antiquities.

Damascus souk, Syria 2010

Jordan remains so far a pleasant and easy place to travel around. With a side trip down into the Dead Sea, we explored new terrain before visiting the classics, Petra and Wadi Rum. A ferry took us to the bureaucratic headaches of entry to Egypt once more. Our route again went south to Aswan, but this time there were no security convoys and no obvious issues with safety.

Camping in the White Desert, Egypt

A big surprise after the crossing of Lake Nasser to Wadi Halfa were the brand-new tarmac roads in northern Sudan. Of course the hassles of the ferry to Wadi Halfa were not a big surprise, although the system overseen by Mr Salah was easier this time, as he remembered us.

Driving along the Nile on a tarmac road seemed a dream after the travails of the railway line route. In no time we were encamped at the Blue Nile Sailing Club, where Kitchener's boat was showing a little more rust.

Khartoum centre, Sudan

Again we headed to Ethiopia and took the quick road to Addis via the Blue Nile Gorge. With some surprising overheating that required four unscheduled stops for tea while the engine cooled down, the climb out of the gorge took hours. In Addis we found Wim's Place, run by a genial Dutchman. It was quite busy with crazy motorcyclists and other overlanding characters. The road to Marsabit in northern Kenya was worse and slower than before. It was a bit scary with incursions by Somali terrorists to the region east of the route.

A big hole in Northern Kenya

The greasy fried egg and toast breakfast at the Jey Jey Hotel was just as memorable as last time. Fortunately a new road built by the Chinese allowed us to get to Isiolo more easily and soon we were back at the Upper Hill campsite and hostel in Nairobi. It was great to be back in the Kenyan capital – old friends, good food and comforts.

Female student graduates in Kampala, Uganda

On this trip we headed to Uganda, now becoming a popular place for backpackers, with rafting and adventure sports at Jinja to enjoy. Since the days of Idi Amin when Bob drove from border to border in one day in 1978, the country has changed radically for the better. Rwanda also seemed to have come a long way since Bob's last visit in 1980, some time before the genocide. In the capital Kigali we camped at the well-named One Love Guest House. The roads were good, the people helpful and the Hotel Foucan in Butare a great colonial remnant.

Motorcycle taxis in Butare, Rwanda

Our new route took us south into Burundi, seemingly stable at that time. We stayed at the Hotel Le Rift in Bujumbura, run by an ex-teacher who had spent some time in Yemen through its earlier troubles. He still seemed to be commuting between Bujumbura and Sanaa. Two more 'odd ball' places you cannot imagine. We were surprised how many mosques there were in Bujumbura.

The countryside south along Lake Tanganyika was stunning. Lush banana plantations, deep green terraced hillsides and canopies of shady trees graced the way. But as the road climbed towards Tanzania, it became equally shocking and exciting, with deep muddy ruts and slopes into ditches that threatened to swallow the Landy. One has to hope that Burundi does not slide off the road into chaos, with a president bent on keeping power. The same old story.

Locals on the Burundi border road

At the Tanzanian border we search in vain for any sign of immigration or customs. The 'road' is a great gash, sliced through layers of exotically coloured mud. A tall bamboo ladder leans against the vertical edge of the chasm. 'Jambo Jambo,' shouts a man above, tottering dangerously close to the edge of the cliff. 'Come on up!' (the price is right!). It was the most pleasant immigration post of the whole trip and no extra charge either in 'local taxes'. 'Welcome to Tanzania! What are you doing here? Are you lost?'

Evidently few foreign vehicles or travellers come through this outpost. Customs was an equally pleasant treat; this time there was no precarious ladder to climb, with steps two feet apart. The drive into Kasulu was OK and we found an enchanting hostelry called God Bless Hotel. Wow, what a revelation this area was turning out to be – such a surprise and so friendly. We had high hopes for the journey to Zambia – foolish – but what the heck.

Since the days of mud paradise in Zaire it's been a rare experience to relive those 'exciting' days. Looking back, the memories are a lot more romantic than the real thing, when we camped knee-deep in thick goo, trying to cook on green wood bombarded by bugs and then being kept awake by the cicadas.

Mud fun in Tanzania

Less mud fun, Tanzania

Next day the road was quite reasonable, not sealed but smooth dirt. After Uvinza, however, conditions deteriorated quickly, with endless mudholes in the semi-wet season. The water-covered holes often have rocks and wood hidden in them, requiring slow, careful driving. Tiny Mpanda boasts the quirky Super City Hotel. Here a strange-looking wood-burning heater dominates the interior courtyard and actually does give off heat – it's surprisingly cold here after rain.

In Katavi National Park, beware the ferocious tsetse flies – it's the only place in Africa where we have ever been in open combat with calculating aggressors like this. In our haste to escape them, the handbrake was accidentally left partially on. A few minutes later, smoke seeping through the floor into the cab indicated that something below us was on fire! Who had to get out and look underneath? Luckily no damage was done, but it does show how hard it is to keep cool at all times. Along the 70km through the park, keep your windows closed! In Sumbawanga we stayed at the relatively luxurious Moravian Mission, and it was with some relief when we pulled into Tunduma, once again on a real road.

Giraffe in Kruger National Park, South Africa

All that remained now was to head through Zambia, Zimbabwe and Botswana to South Africa. We flew back to the UK via Madagascar (now safe) and Nairobi while the Land Rover was sealed up in a small container and packed off to Mumbai by the same people in Durban we had used before. But what lay ahead we could never have imagined.

Tana Palace, Antananarivo, Madagascar

Siân's sad days: May 2010

Although my mother had been suffering from a bad back and was later diagnosed with osteoporosis, I had no idea where it would lead. For some months she had talked about how apprehensive she was about her forthcoming 83rd birthday. Her mother had died of a stroke at the age of 83, and she felt that the same thing was going to happen to her. I arrived back at the end of April, two weeks after her birthday on 14 April, and went straight up to Derby to see her. One evening on the sofa she said, 'I think I'm going to die soon.' I didn't really believe it and tried to ignore what I had heard. Then I decided I should broach the subject again and ask her why she felt that way. 'I don't know,' she said. 'I always thought I was going to live to a ripe old age but now I know I'm not. But never mind, I can't do anything about it.'

Sure enough, on 3 May she had an extreme attack of atrial fibrillation, spent the day in hospital and was sent home on warfarin. A few days after that, on 8 May she suffered a stroke, started to recover, then caught pneumonia and died on 21 May. It was all so sudden, yet every hour seemed like forever. I stayed with her every day in the hospital, putting off our flight to Mumbai twice, as the Land Rover had still not arrived. But eventually I had to confirm our flight to India because the Indian customs would not hold on to the vehicle indefinitely and our work in the Alps was looming within one month. It was an unbelievably stressful time, but she died the night before I had to leave. Did she know? I still wonder how she knew her death was so imminent, what premonition she had.

And should I worry if I ever reach the grand old age of 83?

A boiling Indian Odyssey: May 2010

We were both still numb with shock when we met at the airport for our flight on 22 May. Fortunately our spare clutch and heater unit were not detected at Mumbai airport. Being May, the heat was on and after a few nerve-wracking days at the port the vehicle was released. With little time left before our summer season in the Alps, we took a direct route through Nasik and Dhulia. The heat was on in all ways; the engine was overheating in the 45°C temperatures, and I had to pour water over a tatty white sheet on Bob's head as he drove. A slight detour from Bhopal to Sanchi filled in the historic sight missed on the 1984 Hann trip. Every night Bob worked on the Land Rover, putting more and more aluminium cowling around the radiator fan; enough to get to Khajuraho without overheating. On to Varanasi and the glorious comforts of the Surya Hotel garden, where a loud and lavish wedding entertained us for hours, hours and hours. Now only lovely old Gorakhpur stood between us and the cool of the foothills of Nepal. Climbing to Kathmandu took more hours, as the engine came close to overheating again. But in the end we made it.

Arrival at the Kathmandu Guest House, June 2010

On the day of my mother's funeral we went to the monastery at Boudhanath and had a special personal ceremony for her at exactly the same time. We then flew back to London on Monday, I went to see my father in Derby on Tuesday, got the train back to Chichester on Wednesday and then drove for 15 hours on Thursday night and Friday to arrive in the Alps for the summer with only a few hours to spare before our first group arrived.

****

Having got the Land Rover to Asia, in the autumn we enjoyed a few weeks driving around Nepal and India, visiting old and new haunts, like quaint Bundi, Kota and Ranthambore National Park for tigers before breakfast. The vehicle was parked in Amritsar at Mrs Bhandari's while we visited family in the UK over Xmas and New Year 2011. Before returning to Nepal for our big trip back to Europe overland, we visited Rishikesh and some of the hill stations of north India: Nainital, Mussorie and Dehra Dun.

The delights of Hotel Bissau Palace, Jaipur, India

Bandits, paradise and a canal: January 2011

With earlier trips to Central America dogged by civil wars in El Salvador and then Nicaragua, there were still a couple of countries we hankered to see. Another short trip in 2011 took in Nicaragua, Costa Rica and Panama.

Arriving in Managua, all the guidebook warnings seem to come into play at the guesthouse, which was firmly locked by two lots of iron-barred gates, day and night. Needless to say we saw little of the town. However, further south, around Lago de Nicaragua a totally different atmosphere prevails. Granada is a fine colonial gem and the placid island of Ometepe hosts a picture-postcard volcano and tropical paradise.

Lake Ometepe, Nicaragua

Volcano Poas, Costa Rica

Costa Rica has established a thriving adventure sport industry; its volcanoes and beaches provide enough stimulus for any adventurer. There were a lot of overseas gap year travellers, language students and young teachers in the capital San Jose. The living is easy and the buses don't seem to break down as much as elsewhere.

The canal is an unmissable sight in Panama, but perhaps the sound of gunfire after dark in a less savoury area of cheap hotels in the capital isn't so appealing. Luckily we had a TV with CNN to entertain us. The news was covering the start of the uprising in Tahrir Square in Cairo, Egypt – every hour!

That said, the old part of Panama City, Casco Viejo, hosts some interesting street life and historic buildings.

The Panama Canal

Asia Overland through Tibet: 2011

We could never have dreamt that we would be able to enter Tibet from Nepal. This was a fantasy dream come true. It was the grand finale perhaps; that one trip of a lifetime… Tibet has been forbidden for most of its history and reaching the forbidden city of Lhasa almost impossible. Even now, Tibet is frequently closed to visitors.

In early March we were in Kathmandu preparing the vehicle, putting the heater radiator in for the expected cold weather. Our Chinese permits for Tibet were proving a stumbling block. It could have been anticipated that the paperwork would be more of an issue than the 5000m passes. There appeared to be one paper that needed attention and it later required us to get some photocopies with the seal in red. This, of course, was difficult to get in Kathmandu, where photocopy machines in local shops barely had a serviceable black cartridge let alone colour.

14 April 2011
We finally crossed into China at Zhangmu and were met by 'Penny', our pleasant Chinese guide. We needed a guide to get through the Himalayan paperwork at the border and she was to come with us in the vehicle all the way. It would have been my mother's 84th birthday today, an auspicious day.

15 April
From Zhangmu the road climbs rapidly on to the Tibetan plateau, between the 7000m peaks of Jugal Himal and Gauri Shankar, through the deep gorge of the Bhote Kosi River. It's the same road we hiked down back in 1985. Chilly Nyalam (1hr from Zhangmu) is the first Tibetan settlement, where we stopped to aid acclimatisation to the altitude. 'Beware of Yaks!' says the sign on the road. When yaks ahead bar your road, you know you've reached an overlanders' Shangri-La.

Most of Tibet lies above 4000m, with formidable obstacles to entry placed in the way by nature. Only one road currently crosses the Himalayas into China and it's just 60km east of the Nepalese capital. Kathmandu lies in a lush, semitropical verdant valley but across the mountains the plateau is barren, harsh, wild and unforgiving – a truly exotic contrast from the soft light of the Nepalese foothills. How would our faithful old Landy cope with this amazing physical and geographical step up?

We initially expected some problems concerning the effects of the high passes, not just on ourselves, but also on the aging Land Rover. Would it cope with the 5000m (16,400ft) passes running on possibly badly refined diesel? Can you believe we even kept some Indian diesel from the high country around Nainital for this purpose, thinking it would be better than Nepalese fuel – it probably was. We worried about the bitter cold – would it prove too much for our vehicle? Could we drive at very high altitude, at times only 3000m below the 8000m 'Death Zone'? The ever-industrious Chinese have recently sealed the road. Much to our surprise, the vehicle coped admirably with the high altitude; hardly any loss of power was noticed. In fact, as the air cooled down, the vehicle seemed to gain in strength. The draughty old doors of our ancient steed, however, kept us awake with an icy blast.

16 April
Leaving Nyalam (3750m), we climb relentlessly, passing the cave of Tibetan poet-philosopher Milarepa, to the Lalung La pass. At 5200m it has a breathtaking panorama of the glittering Himalaya giants of Gauri Shankar, Menlungtse and Shishapangma. On this side of the Himalaya the mountains seem close and inviting. From Tingri village there are superb views of Mount Everest and Cho Oyu. The appearance of our Land Rover in their midst provided the local village people with immense fun as they gathered around to examine it in minute detail – the foreign invader from Solihull!

We stayed in grotty lodgings near the old citadel monastery of Xegar Dzong (4350m) for the night. The 'hotel' had no running water, outside corridor loos, draughty windows and strange visitors in the night. The Tibetan landscapes are truly 'other-worldly', with vast vistas and never-ending mountain peaks floating on the horizon.

From Xegar to Xigatse (3900m) the road climbs to a pass, with a great view of Everest before the summit. Lhatse is a further two hours. More small passes and undulating barren countryside leads on to Xigatse. Here is the enormous Tashi Lhunpo monastery, but before sightseeing, to our surprise, we spent most of the time doing paperwork. We needed to have a Chinese MOT-style check on the vehicle before being given our road travel permits, a new number plate and Chinese driving licences in order to continue to Lhasa.

Our Land Rover in Xegar Dzong, Tibet

19 April
Our route from Xigatse to Lhasa was via Gyangtse (3950m) – an enchanting Tibetan town with a great monastery and towering citadel fortress (2hrs drive). After Gyangtse, the road climbs into a defile and continues steeply up, passing a glacier on its way to the 5010m Karo La pass. This is where Sir Francis Younghusband's British military expedition fought the poorly armed Tibetans during their march to Lhasa in 1903–4. It is a 20km descent to the lakeshore and the small town of Ngartze is a few kilometres north around the lake (4–5hrs drive from Xigatse).

Next day, the views around the turquoise blue Yamdrok Lake and from the formidable Kamba La pass (4900m) are truly sensational. From the pass there are more fabulous vistas – westwards over the shimmering lake to the gleaming white Himalaya, while to the north lies a landscape of dark, jagged peaks. Far below and ahead is the silvery ribbon of the Tsangpo River, the gateway to Lhasa. It's a very steep descent from the Kamba La to the Tsangpo River valley that eventually heads into the Kyichu valley.

Lhasa begins at the Yangbajain turnoff, where there are now two main roads into town. Despite massive construction, the city is incredible, and nothing, absolutely nothing, can diminish the impact of the towering Potala.

On the Kamba La above Lake Yamdrok, Tibet

20 April: Into Lhasa

Despite massive modernisation the spirit of old Lhasa remains; the simply stunning grandeur of the former residence of the Dalai Lama, the Potala Palace, still dominates the town. Equally atmospheric is the magical Jokhang temple, where hordes of Tibetan pilgrims gather at the holiest shrine of Tibet. The noises, smells and animations of the Tibetan traders in the Barkhor markets are memorable. The great monastic cities of Drepung, Sera and Ganden are unforgettable; these vast monk temple cities still resonate with monks and stirring prayers. Modernity is all around but delve into a quieter lane and you can still find the mystical, spiritual qualities of old Lhasa and Tibet.

Ganden Monastery, Lhasa

Meeting the locals below the Nyenchentangla peaks

22 April
We head north for Yangbajain, passing the road to the famed Tsurphu monastery. The road heads northeast below the Nyenchentangla Peaks to Damzung, where we stop at a cosy truckstop café for oodles of noodles. Some local Chinese are so taken with the Land Rover they sit on the bonnet for a series of selfie photos. It takes 7hrs to Naqchu.

23 April
From Naqchu to Tuo Tuo there were some strange speed restrictions to adhere to – only 70km per hour – so we needed to wait on the roadside together with the truckers to comply. It's calculated from your time between checkposts. Amdo is a grubby place. There is a steady climb to the Tanggula La pass (5180m). The snows of Shangri-La are common and it's bitterly cold for much of the year. A truckstop town, Yangsping, is reached 60km after the pass. The road continues across the bleak country to Tuo Tuo River town (8hrs drive today). The road wasn't particularly busy, although quite a few big trucks use it.

Morning oil checks in Tuo Tuo, Tibet

The lodge in Tuo Tuo is so new that the TV sets are still in their boxes. We awake to a freezing room and fresh snow. At least we are up-to-date with the news from China CCTV: a new high rise in Shanghai was opened yesterday. As usual the gearbox needs oil, so the cardboard from the TV packaging prevents the icy snow from attacking Bob's back, even if the snowflakes still swirl around. The spanner is a terror to touch, sticking to skin exposed through the holes in the tatty Nepalese woolly gloves. After self-help muesli, tea (not Tibetan), filtered coffee and a few minutes in the icy outside 'bathroom', it's off to Golmud.

24 April
The Feng Liang Pass and Wu Daliang Pass are not the best places to be in April. Visibility is only 3m; the horizon is lost, the side of the road has disappeared and heavily laden, smoky trucks are almost sliding off the edges. Who said Tibet is the last Shangri-La? On this day Shangri-La could have been lost forever beyond the horizon!

Somehow we negotiated the skating rinks of the road in the lowest gear. Even the gearbox oil refused to leak out in this bitter arctic swirl. It was fortunate that we had smuggled a new heater radiator into the subcontinent in anticipation of this 'new experience'. In all the truck-slipping kerfuffle, Siân's not-so-smart phone slipped out into the mist, lost forever on the Tibetan plateau for the Buddhist demons to check out her ancient texts.

Qinghai white-out en route to Golmud, China

Descending now, we left the snows of Shangri-La as quickly as we'd engaged them. It's a dramatic descent through the dry canyons of the Kunlun mountains, surrounded by eroded cliffs and gaunt peaks. Then bliss – the fast run across the plains and on to Golmud for takeaway noodles and Earl Grey tea. That British kettle was priceless in Tibet, despite the funny plugs.

Dunhuang Buddhist caves, Gansu, China

25 April

From Golmud a new highway leads to Ta Tsaidam, with the countryside more desert-like. There were three passes of close to 4000m to cross before reaching the new town of Aksai with some glittering mosques. The Silk Route city of Dunhuang is ahead through some interesting dunes. It's fabulous seventh-century Buddhist art at the Magao cave complex and the high dunes of the Crescent Moon lake area are sites to savour.

A rough but sealed road leads north for 2hrs from Dunhuang to Louliang (willow town). Close to town an engine mounting gives way and, as luck would have it, we get it fixed by some local road construction welders. Unlike the police, whom our guide phones for help first, we get it done for free. The official wanted to charge $300 for a 10km tow to town. We leave a nice tip for the welders and head off with only a bit of the plastic radiator fan damaged – a lucky escape.

It's a long slog to Hami, a town famed for its melons. We take an OK motorway to Turfan but the drive below sea level gets hot, dusty and windy. Turfan has an impressive mosque and a few remnants of the old Uighur town. The next day a good motorway runs through the Tien Shan mountains, with views of Mount Bogda on the right, to Urumchi. The town is unrecognisable. Since our last visit in 1985, it has become a sea of high-rise blocks with flashy restaurants, supermarkets, malls and modernity at every corner. We stay at a Chinese youth hostel full of young Chinese travellers, as excitable as we were thirty years ago, and watch the Royal Wedding of William and Kate.

1 May

Our route crosses the brown soils of the Dzungar plains, with the distant Altay ranges to the north and the peaks of the Tien Shan to the south. It's a good dual carriageway road to Shihezi and on to Jing He. The hotel is great and we enjoy a good rest.

After Jing He is the Samui Lake; the area is beautiful, with mountains, lakes and nomadic herder's yurts. Before the border at Khorgos there is a dramatic drop down from the barren Tien Shan. Because of the May Day holidays in China, we are marooned in Khorgos for three days waiting for exit papers from China. Khorgos is another modern Chinese town but we find an indigenous, Turkic-speaking Uighur historic area to the south of town. They are more than happy to do a minor welding job for us, unlike the hotel people who want to charge us for a towel that is a little dirty.

Finally we are out of China; say no more. The border is easy and changing money is the only issue. A very bumpy road leads to Zharkent, and we are glad not to have our third person in the cramped front of the vehicle.

From Zharkent it's a long drive to Almaty via Shankyn and the Lower Charyn Gorge. It's a quiet desert route with gorges and mountains. We missed the Charyn Gorge; with all the delays caused by Chinese paperwork, we are running out of visa time in Kazakhstan already.

6 May

Almaty is a Russian and European-style city, with some classic museums and stylish buildings. The impressive Cathedral of the Holy Ascension is a colourful highlight here, especially since it is constructed mostly of wood. Getting insurance was easy and very cheap; it can't be of much value other than avoiding any legal hiccups.

After leaving the traffic chaos of Almaty, the road west is good for a while. To avoid the Kyrgyzstan border, a bad and bumpy detour loops around the forbidden 20km or so, taking hours extra via Shu (Chu) to Merke. Once back to the westerly main road, we endure road surfaces that range from awful to fair as far as Taraz.

Taraz is a typical Soviet-style town, with attitudes to match in some quarters. A very bumpy road continues to Shymkent, a historic town dating back to the heyday of the Silk Route, but our time is getting very tight with that un-extendable visa. We are on the road to Turkestan eventually, but finding the right way is often confusing with only Cyrillic signs to guess at. Siân's 'O' Level Russian is proving extremely useful.

Almaty Cathedral, Kazakhstan

8 May
Turkestan is a great place and one of the few in the country to display the rich tapestry of Islamic building. The UNESCO world heritage site, Khodja Ahmed Yassaui mausoleum complex, is a stunning sight; a vision in blue tiling and brown brickwork that matches any in Iran. It has a massive metal cauldron that dates from 1399 and the time of Timur (Tamerlane). The eminent scholar, Yassaui, was born in nearby Sayram and studied in Bukhara, the great centre of learning in Central Asia.

Rab Beg Mausoleum, Turkestan, Kazakhstan

9 May
Not long after leaving Turkestan we stop at Sauran to see the historic mud remains. At one time Sauran was the largest settlement in the region of Kazakhstan, being the capital of the 'White Horde' Mongolian ruler, Sasibuqua. Extraordinarily we met a very British group of keen historians examining the sight with extraordinary enthusiasm for bits of mud remains. The bad, bumpy road continues to the ghostly industrial city of Kyzlorda. This forlorn place is another confusing town and it was hard to find the road out. Extracted from the industrial waste of Kyzlorda, we later cross the Syra Darya River and join the 'main road' to Khosaly.

Cranes on the shore of the Aral Sea, Aral

10 May

Somewhere here is the Baikonur Cosmodrome, with only a slight hint of its location given by a few radar dishes very far away. Before the town of Aral (Aralsk), there are some fleeting glimpses of the Aral Sea. The old harbour at Aral is a sight no one can miss; dried-up, bleak and decaying.

Even the old Soviet-era and once grand hotel seems to be high and dry, with cracked concrete, crumbling staircases and rusty toilets.

From Aral to Aktope there will soon be a new road via Karabuk, but for us it's a stop-start affair with short bits of tarmac and big gaps of dirt. Aktope is a modern town with high rises, busy streets and chaotic traffic. The road from Aktope to Oral (Uralsk) is surprisingly bad. The area of old Uralsk was a big surprise, with grandiose public buildings, churches and theatres. Reminiscent of a bygone era, the buildings are all restored and speak of a Soviet past of pomp and culture.

From Oral a terrible road ensues as far as the border; it's bumpy, potholed and feels very remote. The border is a surprisingly friendly spot, if a bit desolate.

The grand style of Oral (Uralsk), Kazakhstan

13 May

We are on the road to Saratov, in Russia now, and the flavour is much more European. Again the classy buildings in the historic part of Saratov are a surprise to us.

Saratov also has some striking mansions and old streets, as well as impressive onion-domed churches. Of course the rest is drab, dreary and depressing. The endless steppes of Russia are hardly inspiring, nor are the roads; most are surprisingly shocking.

Unfortunately we now only have four days to get across Russia, since our visa time here is also clocking by. They gave us a 10-day transit visa in Kathmandu, but it's not extendable.

The old area of Saratov, Russia

The motorway to Voronezh, Russia

The roads continue the same to Voronezh and even the dual carriageway 'motorway' sections are poor and bumpy. It's just two bad roads side by side. Russian truck drivers ignore all the deep potholes and barrel along like Australian road-trains.

15 May
It takes 7hrs to reach a town called Borisoglebsk. There is only one hotel, but it's a long process to check in, as we are not really authorised to stay in a local establishment. In fact dinner is cooked and eaten in the Land Rover long before the papers are in order. It's a quirky place, with a soft mattress, heavy drapes and creaking doors and floors. Voronezh is another interesting town, but finding a way through it is an epic only slightly less confusing than Saratov. There are more onion-

domed churches and a lot of traffic. Just as we are getting fed up with the confusing signs, we find a cosy roadside restaurant with superb coffee.

Views near Belgorod, Russia

The route continues west to Stary Oskol, an awful-looking town really. Finally we get to the massive city of Belgorod but daren't go into the centre for fear of getting totally lost. Just before the Ukraine border there is a nice new motel that beckons a stay. The menu has some appetising food so we order steak and chips. It seems to be very cheap, until the bill comes and we find we are paying for the meat by weight.

16 May
The crossing into Ukraine is done with only a small delay. It is the first day since leaving Kathmandu that we are not herded about or restricted by visa timeframes. Phew! What a crazy world! Is this what we once understood as the 'freedom of the road'? We'd better go to the USA on a Harley Davidson to enjoy the real freedom of the road, perhaps.

Ukraine is a holiday, with some superb places to visit and excellent food. We drive via Kharkiv, which has a ring road that avoids the big city. Poltava has an interesting heart, with a grand circle of impressive buildings and churches. The road is reasonable to Kiev, with some dual carriageway sections.

18 May
Kiev is bursting with superb churches, monasteries and fine gardens. The Lavra complex is one of those do-not-miss sights. The whole day is one of history, food and relaxing.

From Kiev our route took us to pretty Medivizhy, then south to Kamyanets. The massive castle here is very impressive. To the south is the equally picturesque Kotyn castle. We head on through Chernivitsi to Lviv, via Yeremche and Ivano. Lviv is an historic city with many attractions well worth a day's stopover. The road to the Polish border is not great as yet, nor is the fact that we can't get insurance at the border even though we are finally back in the European Union.

Luckily in Krakow we are able to email a company in London run by Australians for a quote and cover over the phone. Both Krakow and Wroclaw are fascinating cities to visit, and soon we are in Berlin. We visit Hamlin, to see if the Pied Piper is helping with the traffic directions, and soon enough we are exploring the windmills of Holland. In another day our journey is at an end beside the white cliffs of Dover.

Kamyanets castle, Ukraine / The White Cliffs of Dover, England

It only remained to get to Chichester and later a lot of paperwork to 're-register' the vehicle after being away for so long.

So how much did it cost?
The following costs include most visas, insurance, border fees, accommodation (mostly camping) and food

Overland Asia to Kathmandu 1974 (£813 / 3)	£271
Overland Africa to Cape Town 2004	£3030
Overland Africa to Durban 2010	£4700
Overland Asia Nepal to UK via China/Tibet	£3132 (excluding Chinese permits £4000)

Another dream, another adventure

One of our last offerings for this marathon selection of travellers' tales is a short one that had not been possible for virtually the entire period of our time on the road.

Off the map in Somaliland: January 2014

Following a failed attempt to go to Mogadishu in 1975 by Bob, it was not until January 2014 that we finally made it to Somaliland...

Unrecognised internationally as a separate state, Somaliland has functioned quite impressively as an independent democratic entity for over twenty years. In general, with some exceptions, Somaliland has been much safer than the south and it is possible to visit Hargeisa, Berbera, Burao and Shekh with some confidence in the local security services. That said, it still rates as a troubled and risky place with terror threats/kidnapping on most travel advisory websites. The visas were readily and eagerly given to us by the Somaliland Mission in London.

From Addis Ababa we flew into Hargeisa with some trepidation. Also on the plane were some British immigration officers being sent to train their Somaliland counterparts. We took a taxi to the friendly and welcoming Hotel Oriental, who immediately put us at ease. However, we did require an armed escort for our 2-day tour.

The bustling markets of Hargeisa are as colourful as any in Africa. It was apparently safe enough to walk around and a lot of curiosity came our way. We knew that a steady stream of travellers had already taken this path, but that doesn't mean it's a good idea. The only hassle came when we tried to photograph the beautiful downtown mosque; this, it seemed, was one thing to avoid. Photography was best restricted to through the closed windows of the hotel or car.

Lady in Hargeisa market, Somaliland

In the morning we departed with Lucy, an Australian, as company on our tour. She had already hitched to Hargeisa from Djibouti, an amazingly brave or foolhardy venture, depending on your point of view. Las Geel was a true revelation; its exotically shaped outcrops shelter stunning artwork dating from maybe 10,000 years ago. Brightly painted cows clearly illustrate the pastoral lifestyle of that time.

Las Geel ancient rock art & our guard

Sleepy Berbera was a surprise, its old quarter full of crumbling Ottoman buildings and wide sandy streets. Despite being a historic port of some note, there is not yet much of a new quarter.

Street in Berbera, Somaliland

Hargeisa, Somaliland

From Hargeisa we took a taxi from the hotel to the border at Tog Wagale with Ethiopia; the road was almost all sealed and probably is by now.

It's a shame that the country has not yet received international recognition, but it is doing very well for itself without any official foreign aid.

Were we adventurous or simply stupid? Sadly there's no way of knowing how safe this revealing experience is.

Into the Cauldron of Fire, Erta Ale: January 2014

Across the border in Ethiopia, the town of Harar is a fascinating place for a stopover. The city walls are intact, the streets narrow and mysterious. Veiled women watch from airless, darkened rooms through tiny windows on to a scene of colour. Exotic wares and Chinese goods are being sold from every corner and alley. There are dozens of mosques, a church and many stylish Muslim buildings. The place seems to have hardly changed since Burton and Rimbaud explored the town in the 19th century.

But we have not come to Ethiopia just to see Harar, although it's long been on our list. We hope to see one of nature's most amazing spectacles. It lies east of Mekele, where the terrible drought and famine of the 1980s were so visually exposed by Michael Burke of the BBC. We join an ad hoc group of budget travellers on a trip to the Danakil arranged locally when demand is sufficient.

Despite a scary brush with armed militia while having a tea stop en route to Djibouti in 2004, we were still keen to return to the Danakil. Taking a 4-day tour from Mekele, our group of nine independent travellers descended into the Rift Valley via Berehile and some impressive canyons to Hamedella. Located close to the vast salt lake of Assale, this scruffy Afar settlement provided little but hot wind, tatty wooden beds and a mobile phone mast, but no toilets or running water. Its best attributes were the colourful camel caravans and a stunning starry sky.

Dallol, 25km to the northeast, is a multicoloured, earth-shattering wonder of nature -- a series of weirdly shaped fluorescent formations, exotically shaped outcrops and smoking chimneys. Green and blue bubbling pools bordered with dazzling white salt crystals are studded with colours in all imaginable hues. Close by is a forest of grey volcanic towers and a large blue pool with bubbling yellow sulphur fountains.

It's a six-hour desert drive from Hamedella followed by a three-hour hike in the dark from El Dom to reach camp on the crater rim of the Erta Ale volcano. No words can describe the sense of awe that is felt on arrival at the top in the darkness of the night. Glowing a fiery red, the lava lake is approached after a short descent from the crater rim into the hopefully dormant cauldron of this hell fire. We walked across contorted lava flows, frozen into brittle serpentine coils and vast tubes, the petrified crust crunching at every step. Standing at the lake rim, only 10m below us the turbulent molten lava, boiling and cracking loudly, exploded periodically into sensational fiery crescendos of white, yellow and red.

Super-heated red jagged lines zigzagged across the treacle-like flows in the eerie darkness of the midnight hour. Mesmerised by the primordial display, two hours passed in a flash (or two!). Before dawn the show began again, as we watched the exploding lava eruptions before the orange glow of dawn. Noxious gases drifted and swirled across, making the picture one of sheer amazement.

Erta Ale lava lake at dawn, Danakil, Ethiopia

Were we to be born again in two or three hundred years, we would probably want to go and explore those distant realms beyond our current imagination. Sailing past the moons of Saturn, brushing past a black hole or switching universes through a wormhole.

Who has not looked up to the heavens and pondered about that which is beyond our reach?

Erta Ale lava glowing at night

Erta Ale is an intoxicating sight, an inspirational place of wonder, as unfathomable as the stars and the universe.

Down Under briefly: Spring 2015

Following a few years of short breaks due to the sad issues of Bob's ageing parents, we made a longer trip to the southern hemisphere in the spring of 2015. With so much flying, it was the most expensive trip we have ever done.

Darwin, missed by Bob in 1974 because of Cyclone Tracy, was our landfall destination in Australia after a brief stopover in Singapore and then Dili, East Timor, also missed so long ago.

We stayed at Dili Backpackers, with very few travellers around. To the east of Dili are some amazing, deserted beaches beyond the statue of Christ on a hilltop. That is the main tourist sight of Dili. The town is a mix of rundown and swish new, funded by foreign aid. Peace seems to be sticking, so the prospects look rosy.

Beaches awaiting development near Dili, East Timor

Uluru outcrop, Central Australia

A three-week driving trip took in the road to Alice Springs, Uluru, the Kata Tjuta outcrops and Kings Canyon. The rest of the journey took us from Darwin to Cairns by plane and then via Port Moresby in Papua New Guinea to Brisbane. We hardly did justice to Papua New Guinea; taking the local buses around Port Moresby was fascinating though not recommended. The people seemed intrigued by the fact that we dared to travel with them even though we didn't know or particularly care where we were going – simply sightseeing. The fabulous resort of Loloata deserves all the praise it seems to get; it's a truly beautiful tranquil place.

Koki beach area, Port Moresby, Papua New Guinea

Cyclone damage on Efate, Vanuatu

Our trip to Vanuatu was totally marred by the worst cyclone the country has ever experienced. Our own disappointment, though, paled into insignificance compared to the conditions for the locals in Port Vila and across the country. Little help seemed to be getting to the rural population on the main island, Efate. We had hoped to visit the spectacular volcano on the island of Ambrym, but it was not possible. We contented ourselves with being lucky to get to the island at all, and spent a pleasant and restful week taking walks out of Port Vila. There was no public transport outside the capital. With such friendly people, the island nation is sure to attract tourists back in the coming dry season.

We flew on to New Caledonia. Noumea proved extremely agreeable, modern, dynamic and blessed with beautiful beaches.

Noumea harbour, New Caledonia

The youth hostel had the best possible location in all of Noumea, on top of a hill (aren't they all?) overlooking the town with a fabulous panorama of the bays. Noumea could be any town on the French Riviera with lots of wealth and modernity, but in the quiet backstreets it's not quite so brash and breezy.

On the local bus to Hienghene we were the only tourists. It's a fabulous drive, with rugged coastlines on both sides of the island. The central mountainous area is lush and wild. The east coast settlements were very quiet and sleepy, including Hienghene. The coast nearby is peppered with incredibly spiky, black outcrops in weird shapes. One is the chicken head rock; they are worth all the effort to get there.

A lot of the rural areas inhabited by the original Melanesian Kanak people, away from the immediate area of Noumea and Bourail, remain very poor.

For French expats it seems like paradise, but how long can France hold on to its Pacific island paradise?

Linderalique outcrops near Hienghene, New Caledonia

Our last night in Noumea was bliss, a fiery red sunset cast spells over the calm balmy water of the harbour. From the hostel's panoramic vantage point, the lights of the city twinkled like a fairytale; it was magical. The next day during our flight to Brisbane another shocking natural disaster unfolded thousands of miles away.

Earthquakes in Nepal: April – May 2015

In 2011 Cicerone commissioned us to write a new guidebook to the Annapurna region of Nepal. Since then we have been working with a local Nepalese publisher, Himalayan Map House in Kathmandu, producing a series of trekking guides that cover all the main trekking areas of Nepal. It's been a superb job, as we never had the opportunity before to trek in the more off-beat regions like Manaslu, Ganesh Himal, Rolwaling and Dolpo. We were all set to head to the western region when our plans were aborted cataclysmically.

In April and May 2015 two powerful earthquakes struck Nepal, causing massive disruption to the country. Although many older houses and some historic temples were left in ruins, across Kathmandu the majority of buildings and infrastructure remained intact. Sadly the rural regions adjacent to the two quakes suffered more serious damage. The main areas affected were below the peaks of Manaslu, Ganesh Himal, Langtang, Gauri Shankar and parts of the Everest region. The entire Annapurna region, including Upper Mustang and Nar-Phu, was miraculously spared any significant damage. Transport links between Kathmandu and Pokhara remained open and functioning normally.

We were in Kathmandu during the month of May and survived the second earthquake on the fourth floor of the Hotel Moonlight. Having becoming instant aid workers (buying up rice, tarpaulins, tin sheets and warm, locally made clothing using generous donations), we witnessed a remarkable few weeks in the country. After the first days of shock, thousands of local people, young and old, engaged in the relief and rebuilding process with amazing energy. Our old and new foreign friends David Durkan, Ian Wall, Frank Lutick and Steven Stamp, along with many others too numerous to mention in this book, also got stuck into the rebuilding process.

See our Earthquake Diaries book for more information. There is no doubt that the resilient people of Nepal will be back on their feet well ahead of expectations.

Earthquake damage, old Kathmandu

The future …

Whether overland travel to either Africa or Asia in future will get any easier is anyone's guess. All we can do is keep on dreaming!

Alas, after so many years it seems to get harder and harder to travel around the world. Nobody has given us this privileged right to expect to go everywhere, but then again does anyone truly have a right to say we can't? Where there is a will, even a citizen of a country like Nepal can find a way to travel. Our friend KC managed it before 1974, travelling on a cruise ship from Bombay as a cook.

We ourselves were lucky to be able to doss down with Bob's parents in a 'granny flat' shed when we stayed in the UK. For most of the earlier years we could get temporary factory work to boost the budget. A lot of our travel was with our jobs in the adventure travel world, but today these jobs are not so easily found. Most countries now have well-qualified local leaders that reduce the choices for foreign guides. Half of our life we have slept in tents, under mosquito nets, in the front and back of trucks and buses or vans in the Alps. We survived in Africa on bread and bananas twice a day and always stayed in the grottiest hotels when necessary. Comfort and food were not our main concern.

Of course we can appreciate that some may view these stories and our lifestyle as frivolous, selfish and pointless. There are far too many people in the world who have no such chance or choice. We make no claims about the morality of our path or why we did it all.

In the end it was probably reasonable health and an overwhelming passion to see and understand as much of the world as possible that carried us on.

And there are always more places to visit, some closer to home.

In Search of the Green-Eyed Yellow Idol

There is a vibrant quality to life in most places we have been, and in some a certain spirituality. For us it is the Kathmandu of the 1980s that we remember and can still find where we often feel most content. We cannot define anything specifically; it's just one of those places that 'holds you down', as the Cat Stevens song says.

*'Kathmandu, I'll soon be seeing you,
and your strange bewildering time
will hold me down.'*

Perhaps we all have a place or places that ignites a hidden spark in the depths of our subconscious?

From the rooftop we will soon see the hill of Pashupatinath. Among the shrines of that revered Hindu place of worship, there must be many lost idols, some buried by the encroaching forest. Down by the river north of the burning ghats is a temple falling into disrepair; the idols here are sacred to both Buddhists and Hindus alike. Monkeys scurry about searching for food scraps, ripples flow across the river as it trickles down to the ghats. A crow sits on the stony-faced idol, pecking at the red vermillion powder. The idol remains aloof; its eyes have seen the apparent madness of humanity all around.

In the heart of old Kathmandu is a shrine encased in metals of all colours. Pigeons gather to observe the rituals performed here since time began. Intricately stamped images of the Lokeshvaras, the disciples of Buddha, clad the lower plinths of the temple. The god who is worshipped by both Buddhists and Hindus sits serenely inside, bedecked in a garland of marigolds. His strange expression gives a hint of another world. He stares out, knowing more than he is willing to let on. He might be from another planet; perhaps he waits to return there. It might be a more serene place. He came long ago from Assam: a mysterious god, a god whose faithful worshippers attend every day, for he brings life to the valley, the Rain God, Machhendranath. Is he the green-eyed yellow idol?

On the hill to the west of Kathmandu is Swayambhunath. Here is the lotus flower that emerged from the ancient mystical lake that once covered the valley. From the forest-clad hill a petal of the lotus flower is said to have floated to Pashupatinath, a divine link of ideas, a mythical link of purity between the ideas and ideals of two interwoven creeds. According to an ancient Hindu text, when the gods have been aroused a dark age will come upon the earth and its people. What is the truth of the long-lost green-eyed yellow idol?

As for us, we will continue to return to the great stupa of Boudhanath for inspiration. The eyes will have been repainted, new flags added and the top repaired after the earthquake. The lower white dome of the stupa will be repainted with yellow loops – the yellow paint that is thrown with great skill from a bucket on an auspicious day. There are certainly more idols than tourists in Kathmandu these days in the wake of the quakes.

Searching for the idol is merely our metaphor for seeking the mysteries of life itself. Most of us may never find it. Perhaps the true essence of life was discovered here under the gaze of the eyes hundreds of years ago by wiser sages than us.

*Om mani padme hum, Om mani padme hum
Peace to the jewel in the lotus*

The eyes look out from Boudhanath stupa

Even after all these adventures, we have still not found the green-eyed yellow idol.

It must be somewhere, lying neglected, where the gods need pacifying and where dreamers and travellers are drawn by an addiction to find their own Shangri-La.

It's the beginning that's the worst,
then the middle,
then the end.
But in the end,
it's the end that's the worst.

Samuel Beckett

Made in the USA
Charleston, SC
14 January 2016